EcoPopulism

Social Movements, Protest, and Contention

Series Editor: Bert Klandermans, Free University, Amsterdam

Associate Editors: Sidney G. Tarrow, Cornell University
Verta A. Taylor, The Ohio State University
Mark Traugott, University of California, Santa Cruz

Volume 1.   Andrew Szasz, *EcoPopulism: Toxic Waste and the Movement for Environmental Justice*

EcoPopulism

Toxic Waste and the Movement for Environmental Justice

Andrew Szasz

Social Movements, Protest, and Contention
Volume 1

University of Minnesota Press
Minneapolis
London

Published by the University of Minnesota Press
2037 University Avenue Southeast, Minneapolis, MN 55455-3092
Printed in the United States of America on acid-free paper

Library of Congress Cataloging-in-Publication Data

Szasz, Andrew, 1947–
    EcoPopulism: toxic waste and the movement for environmental justice / Andrew Szasz.
        p.   cm. — (Social movements, protest, and contention ; v. 1)
    Includes bibliographical references and index.
    ISBN 0-8166-2174-8 (alk. paper)
    ISBN 0-8166-2175-6 (pbk. : alk. paper)
    1. Environmental policy.   2. Environmental protection.   3. Green movement.   I. Title.   II. Series.
GE170.S9   1994
363.7—dc20   93–28970
                                                                 CIP

The University of Minnesota is an
equal-opportunity educator and employer.

To Aaron

Contents

Acknowledgments   ix

1. Introduction: Environmental Crisis and the Search for a Politics That Works   1

Part I. Policy; Icon; Social Movement: Hazardous Waste in Three Arenas of Political Action

2. Routine Regulatory Failure: The Resource Conservation and Recovery Act of 1976   11
3. "Toxic Waste" as Icon: A New Mass Issue Is Born   38
4. The Toxics Movement: From NIMBYism to Radical Environmental Populism   69

Part II. Reactions

5. Could Opposition Be Neutralized? Discourses and Policies of Disempowerment   103
6. Hazardous Waste Regulation Progresses against the Conservative Tide   116

Part III. Results

7. Fifteen Years of Hazardous Waste Legislation: Summing Up the Policy Impacts   137
8. Broader Political Implications? Environmental Populism and the Reconstitution of Progressive Politics   150
9. Concluding Remarks   162

Notes   167
References   195
Index   213

Acknowledgments

Many friends and colleagues contributed to the making of this book. I would like, first, to thank the University of Wisconsin faculty who taught, advised, and encouraged me when, as a graduate student, I first became interested in studying regulatory policy: Erik Olin Wright, Michael Aiken, David Mechanic, the late historian Harvey Goldberg. Colleagues at the University of California, Santa Cruz, Paul Lubeck, Jim O'Connor, Wally Goldfrank, David Goodman, Mark Traugott, and Bob Alford, offered much-appreciated support and encouragement at the other end, as the project neared completion. My research assistant, Hal Aronson, helped me understand the importance of the personal political transformations that were occurring in the toxics movement and conducted interviews with key movement leaders that, with great skill and sensitivity, elicited compelling oral political autobiographies (excerpted at length in Chapters 4 and 8). Over the past several years, I have subjected my colleagues in the American Sociological Association's Environment and Technology Section to various fragments of my findings and arguments. I wish to thank all those who listened, asked stimulating, challenging questions, and expressed their eagerness to see the finished work: Vern Baxter, Stan Black, Riley Dunlap, Judy Friedman, Bill Freudenburg, Lawrence Hamilton, Craig Humphrey, Celene Krauss, Steve Kroll-Smith, Adeline Levine, Judy Perrolle, Tom Rudel, Emily Schmeidler, Alan Schnaiberg, James Short, Peter Yeager, and others.

Even as this list grows, I realize I cannot possibly name everyone who ought to be thanked. It is easiest to recall and list the most proximate influences, the friends and colleagues who were there as the project was conceived and executed. It is harder, though just as appropriate, to honor the others, all those teachers one never meets but whose work shapes how one goes about the work of social research and analysis. To do justice to those more distant, more powerful influences would require, in effect, that I reconstruct the whole of my intellectual development, a task for which, at this moment, I have neither the space nor the will. I will have to trust, instead, that my deepest, most enduring debts will be apparent as the text is read.

Richard Flacks twice reviewed drafts of the manuscript for the University of Minnesota Press; his comments and criticisms proved essential in the final stages of manuscript revision. Lisa Freeman, my editor at the Press, was all that one hopes for in an editor. Tough, rigorous in her expectations, yet unfailingly supportive, she provided exactly the right context as I struggled to produce the final draft.

My wife, Wendy Strimling, a lawyer as well as a sociologist, was immensely helpful. She was usually the first to hear a new thought or argument, and I found her reactions trustworthy and accurate. More specifically, her knowledge of the law proved invaluable to my understanding of the legal aspects of regulatory law and implementation. Finally, I wish to thank her for her tolerance. Researching and writing consume immense amounts of time. Months turn into years. One loses all sense of limiting working time to normal hours. Even if physically present, one finds oneself mentally absent—off in the virtual reality of one's text. Obsessed writers make less than scintillating company.

I am deeply grateful to the activists and movement leaders who generously consented to be interviewed for this book. Chapters 4 and 8 could not have been written without their cooperation. The videotapes of network newscasts that I used in chapter 3 were provided by the Vanderbilt Television News Archive. For the rest of the material I used, the documents and the vast secondary literature on toxic waste, I must thank the University of California library system, especially at the Berkeley campus. I found its holdings in this field, scattered across almost a dozen campus libraries—Law, Engineering, Public Health, Business/Social Science, Environmental Design, Government Documents, and several more—comprehensive, if not overwhelming.

Parts of this research were funded by grants from the Rutgers University Research Council and the University of California, Santa Cruz, Social Science Division. Sections of chapters 2, 6, and 7 have appeared in somewhat different form in *Social Problems, Criminology, Science and Society,* and *Capitalism, Nature, Socialism* (Szasz, 1984; 1986a-c; 1992).

Chapter 1

Introduction:
Environmental Crisis and the Search
for a Politics That Works

"I saw the news today, oh boy"

Every day, through every news source, we are increasingly made aware of a complex and pervasive environmental crisis. In 1989 alone the American public saw and heard the following environmental stories on the television networks' nightly newscasts:[1]

Oil production and consumption: The *Exxon Valdez* oil spill, by itself, guaranteed that the pollution effects of oil would be the biggest environmental story of the year. I start with this, too, because oil lies at the very heart of the economy and the fabric of everyday life. At the production end, the news reported one oil rig on fire, explosion and fire at three refineries, two pipeline explosions, and ten tanker accidents, most of them resulting in oil spills. At the consumption end, urban air pollution was reported to be as serious as ever, and studies increasingly linked carbon dioxide emissions to predictions of planetary climatic change.

Commercial nuclear power: There were reports of new troubles at some plants, updates about past accidents, criticisms of regulators, and reports of elevated rates of cancer and childhood leukemia near plants. At the tail end of the nuclear fuel cycle, a safe permanent way to dispose of nuclear wastes had still not yet been found, spent fuel was accumulating, and temporary storage capacity was growing short.

Nuclear weapons facilities: Commercial nuclear power has been controversial for years; in contrast, society had been, until recently, kept ignorant of the probably more serious environmental and public health threat from improperly managed federal nuclear weapons facilities. In 1989, that veil had begun to part. Safety problems at facilities throughout the nation were reported. Communities adjacent to federal facilities discovered that they had experienced decades of exposure. Serious impacts on the facility employees' health, long ignored or suppressed, were retroactively documented.[2]

Impacts of production, more generally: New federal reporting requirements

1

produced data showing that industrial discharges of toxic pollutants into air, ground, and water are far greater than previously thought. Stories documented continuing improper management of hazardous industrial wastes. Coal-burning electric generating plants cause acid rain. Other stories detailed levels of exposure to toxics in the workplace, recalled earlier industrial accidents, such as the Union Carbide catastrophe at Bhopal, and reported a handful of new catastrophes at various industrial sites.

Agriculture, fishing: In this category, the focus was on pesticides. Some pesticides' capacity to cause cancer was again confirmed. Farm workers were said to be at greatest risk, but images of schoolchildren eating Alar along with their daily apples conveyed that pesticide residues threaten consumers' health as well. Pesticide and fertilizer residue runoff was also shown to be polluting land and water and threatening wildlife. Other stories showed how modern fishing methods threaten ocean mammals and sea turtles, how their very efficiency fatally disrupts whole oceanic ecosystems.

Urban pollution, consumer waste: Cities in California and the Northeast were singled out for having the worst air; reports from the U.S. Environmental Protection Agency (EPA), Congress, and environmental groups documented the dismal state of air quality nationally. Urban wastes polluted beaches, harbors, and oceans. Several cities were found to be in violation of the Clean Water Act. Stories about a freight train full of sludge traveling from place to place and finally returning home after failing to find a facility that would accept the wastes dramatized the growing landfill capacity crisis.

Species: A large number of species were reported to be declining or at risk of extinction owing to destruction of habitat, pollution, overhunting, and poaching. Reports documented the plight of mammals (elephant, walrus, black rhino, whale, dolphin), reptiles (desert tortoise, sea turtle), birds (peregrine falcon, trumpeter swan, pelican, sandhill crane, bald eagle), many species of amphibians, and the giant saguaro cactus. One report noted that, overall, a great extinction was under way.[3]

Geographic areas, ecosystems: Beyond the threat to individual species, large geographic areas and whole ecological systems were at risk. Economic exploitation and development threatened to transform and hence destroy entire ecological systems: rainforests in Asia and Central and South America, remaining old-growth forests on the Pacific Coast, and wetlands across the nation. Whole seas, the Adriatic and the Mediterranean, were reported to be seriously polluted. Wilderness, wildlife refuges, and national forests and parks ranging from Alaska to Antarctica, from the Grand Canyon to the Everglades, were reported polluted by industrial discharges and threatened by encroaching development, logging, livestock grazing, or oil drilling.

Threats on a planetary scale: The evidence grew that the planet and its web of life, as a whole, are at risk as chemicals break down the ozone shield and greenhouse gasses accumulate. The ozone hole at the South Pole was reported to be getting bigger, and new studies documented ozone depletion at the North Pole as well. Governments, manufacturers, and environmental organizations all agreed that chlorofluorocarbon emissions must be decreased, but effective international action seemed far away. Scientists had predicted that carbon dioxide emissions might cause significant global warming; now, widespread drought and record heat were being interpreted as its first observable manifestations. But, reports said, action would be costly for the U.S. economy, and the Bush administration was resisting pressure to sign international agreements.

Taken as a whole, these stories documented a crisis of extraordinary depth and breadth. We may not know, even now, the full extent of the damage that has accumulated over the past century. What we already know is more than enough to make thoughtful persons cringe. We cannot help but be filled with concern for ourselves, for other creatures, for our children and our children's children. As a world society, we appear to be fast approaching an ill-defined yet palpable moment of criticality. It becomes ever clearer that continuing down this same road will lead to certain ruin for both the planet's web of life and human civilization.

The Two Faces of Modernity

The contradictions of modernity were evident to thoughtful observers from the very first. On the one hand, it was abundantly clear that capitalism had remarkable, unprecedented capacity to produce material wealth. Even the most unrelenting critic had to acknowledge its productive powers. In the *Communist Manifesto*, for example, Karl Marx marveled that in "scarce one hundred years, [capitalism had created] wonders far surpassing Egyptian pyramids, Roman aqueducts, and Gothic cathedrals" (in Tucker, 1978:477), that it had created "more massive and more colossal productive forces than all preceding generations together" (ibid.:476). Since then, capitalist society's generative powers have proven, if anything, even more impressive. It has brought us, at least some of us, in some of the world's nations, wonderful things: a high standard of living, longevity, technological marvels, ready access to vast fields of cultural goods, and unprecedented personal mobility.

On the other hand, it was also evident to observers in the nineteenth century that unfettered economic development came with a large price tag in terms of human suffering. Journalists, physicians, government inspectors, novelists, the first great sociological theorists, even the political economists who wholeheartedly embraced the newly emerging system—all, in their own way, described the human consequences, the destruction of traditional social forms, dislocation, mass impoverishment, and widespread hunger and misery.[4]

Other adverse consequences were, at the time, less clear. Discussion of potential impacts on nature was limited to highly visible impacts that had immediate, obvious consequences—soil erosion that threatened to compromise future food supplies, for example, and concern about modern production's impact on workers' health.[5]

A century later, the social justice consequences are as pressing as ever, but our concerns about them are now equaled by concern about adverse impacts on nature. We now understand as never before that the cornucopia of modern development has been accompanied by, premised upon, and made possible by the increasingly destructive transformation of Earth's ecological, even climatological, systems.

Searching for the Way Out

By now, of course, practically everyone recognizes the gravity of the situation. The shelves are filling up with books that tell us point blank that we face extinction and that we will take many of God's creatures with us as we go. We are told to transcend anthropocentric consciousness, rethink materialistic values, totally reorganize the built environment and the patterns of everyday life. Environmental scientists warn that we may have no more than forty years in which to make these radical changes.[6]

But how? How do we move toward a *sustainable civilization*, toward a reconciliation of human activity with the planet's capacity? To call for fundamental restructuring of production is to threaten the interests of society's most powerful elites. Even modest reforms threaten someone's livelihood and are therefore resisted. Much of the ongoing damage is inscribed in the historically sedimented accumulation of ways of living—the car, housing, urban and suburban geography, patterns of consumption—whose sheer facticity and inertia promise to overwhelm conscious efforts to change any of it.

There is, certainly, value in depicting as clearly as possible the place we must reach, but it only deepens our despair when we hear calls for fundamental change without realistic discussion of how that change could be achieved. We are sorely in need of plausible scenarios for moving forward.

Hope lies in two facts. First, we know better than ever the situation we face. Second, environmental politics is in creative ferment. While mainstream environmental organizations continue the traditional fight for stronger regulatory legislation, other, newer voices in the environmental movement are opening the movement's tactical options, proposing an extraordinarily wide range of reform scenarios. Some advocate the politics of the exemplary act: direct, militant action. Others have opted for a more traditional social movement form, community organizing. Still others advocate individual, voluntary behavioral change, buying green and recycling. From fifty simple things you can do in your own home, in your spare time, to save the Earth, to Earth Summits where nation-states try to forge international agreements, from ultraleft "ecotage" to market mechanisms such as the buying and selling of pollution rights, almost every imaginable solution has its advocates.

All this, together, constitutes a vast laboratory of practical political experimentation. Some, perhaps most, of these new strategies and tactics will prove ineffective. One or more might come to be seen, in retrospect, to have been the first concrete step toward the generation of a politics that can resolve the crisis we have entered.

Have there been some successes already in this new, more eclectic phase of environmentalism? Are there instances where real steps have been taken? If so, we ought to study them carefully, understand what has worked and why.

"Toxic Waste": A Case of Success

I would suggest that hazardous waste politics and policy constitute one such case of success. The issue of toxic, hazardous industrial wastes has been arguably the most dynamic environmental issue of the past two decades. As recently as 1976, "toxic waste" was not yet a well-formed social issue. There was no clear public opinion concerning it, no crystallized mass perception that it is a serious threat to people's health. Hazardous waste became a true mass issue between 1978 and 1980, when sustained media coverage made *Love Canal* and *toxic waste* household words. By 1980, the American public feared toxic waste as much as it feared nuclear power after Three Mile Island.

The overall political tenor of the 1980s was as conservative as that in any period in recent memory. Labor, the traditional core of the progressive movement, suffered defeat after defeat. Its once-formidable political clout declined precipitously. The New Left was in disarray. The heady, optimistic activism of the sixties was a thing of the past. The populace seemed split between increasing indifference to politics and increasing acceptance of conservative positions. Movements struggled to stay alive in these inauspicious circumstances. In the policy realm, the Reagan administration was bent on undoing social legislation dating not only from the sixties, but from the New Deal. Champions of deregulation, administration officials assaulted the environmental laws that were passed around 1970. New regulatory initiatives died quickly and quietly in congressional committees.

Given that context, the post-1980 development of the hazardous waste issue is all the more remarkable. While other movements struggled just to stay alive, concern about toxic industrial waste sparked a widespread, dynamic social movement. Thousands of local, community-based groups formed. In less than a decade, a rich infrastructure of more permanent social movement organizations appeared.

At the beginning, the movement's earliest groupings had little that could really be called an ideology. They acted out of a relatively narrow, simple NIMBY ("not in my backyard") consciousness. From those humble beginnings, impressive things grew. The movement expanded its focus to include other local contamination issues, municipal waste, military toxics, pesticides, and others; it was no longer just a "hazardous waste" movement but a "toxics" movement. At the same time, the movement's understanding of the toxics problem deepened, became more systematic and more radical. The movement's combination of radical critique and direct, grass-roots tactics is perhaps best described as "radical environmental populism."

The facts that "toxic waste" inspires extraordinary levels of fear and dread and that people are evidently more than willing to get involved in fighting its spread did not escape politicians' attention. Consequently, the development of hazardous waste policy has been very dynamic. The federal government first regulated industrial waste in 1976, almost incidentally, as a minor and uncontroversial addition to legislation dealing with several barely related concerns: resource conservation, resource recovery, and household waste. After 1980, even as other environmental regulatory efforts stalled, lawmakers responded to the sudden rise of public concern and the surge of social movement organizing by strengthening both major hazardous waste laws when they came up for reauthorization. By the end of the decade, hazardous waste policy was moving toward a break with the traditional regulatory format of pollution *control*, toward the superior logic of pollution *prevention*.

Impressive as these policy impacts were, to focus on them alone would be to miss other and potentially more important impacts. The movement was the cauldron in which a new kind of environmentalism—what I am calling *radical environmental populism*—formed. The movement brought a whole new mass base of working people and people of color to environmentalism. It forged practical and conceptual links between environmentalism and the struggles against racism and sexism. Most recently, it has articulated the position that environmentalism is not just one more issue that exists alongside, but unconnected to, the other great social causes of the day. Rather,

> There is an unbreakable link between the environmental issue and all the other troublesome political issues. . . . environmentalism reaches a common ground with all the other movements . . . civil rights, women's rights, gay and lesbian rights, anti-war, against nuclear power and for solar energy, world peace, . . . the much older labor movement. (Barry Commoner, *RACHEL's Hazardous Waste News*, 30)[7]

> The new Grassroots Environmental Justice Movement seeks common ground with low-income and minority communities, with organized workers, with churches and with all others who stand for freedom and equality. (Citizen's Clearinghouse for Hazardous Wastes, *Everyone's Backyard*, 1990, 8[1]:2)

Some of the more thoughtful voices in the movement have begun to suggest that environmental populism is not only the vehicle for achieving a better environment, but also the most promising candidate for leading the next phase of the larger movement for social change.

How did all this happen? How did a nonissue so quickly become a passionate ISSUE, "the nation's most important environmental problem"? How, why, under what conditions did perceptions so easily become radical action? How did action informed by rather narrow and even apolitical self-interest (NIMBYism) generate one of the most radical environmental ideologies we have today? What impacts has the new environmentalism had on policy and, more broadly, progressive politics? What can we learn from the history of hazardous waste as we search for a way out of this crisis and toward the necessary reconciliation of nature and human activity? These are the questions that I address and hope to answer in this work.

Part I

Policy; Icon; Social Movement: Hazardous Waste in Three Arenas of Political Action

Chapter 2

Routine Regulatory Failure:
The Resource Conservation and
Recovery Act of 1976

Modern industrial production generates copious quantities of toxic by-products. In 1989, for example, American companies reported to the EPA that they had released 5.7 billion pounds of 325 highly toxic substances into the environment. That same year, the American Chemical Society reported that the nation's firms had generated somewhere between 580 million and 2.9 billion *tons* of hazardous wastes.[1]

As late as the mid-1970s, the disposal of these often highly toxic materials was almost totally unregulated. Federal clean air and water statutes provided some "authority over the incineration, and water and ocean disposal of certain hazardous wastes"; fourteen other laws dealt with various aspects of hazardous waste management "in a peripheral manner" (EPA, 1974:ix). No federal law regulated land disposal, by far the most prevalent form of industrial waste disposal. Federal regulation was, then, woefully incomplete, and what little existed was highly fragmented. Things were no better at the state level. Twenty-five states had enacted some legislation, but, as the EPA reported to Congress in 1974, "hazardous wastes . . . are essentially unregulated in practice, for none of the 25 jurisdictions has fully implemented its control legislation" (ibid.:17).

This era of nonintervention ended in 1976. In passing the Resource Conservation and Recovery Act (RCRA),[2] Congress created the first comprehensive national program to regulate the treatment and disposal of industrial hazardous wastes.

Ordinarily, legislative action is stimulated by and follows issue creation, that is, the moment when some condition comes to be widely perceived as an alarming "problem." In the case of RCRA, however, this familiar sequence was reversed. "Hazardous waste" did not become the compelling political issue it is today until 1978, well after congressional passage of regulatory legislation.

The undeveloped state of the issue skewed both the legislative process and the final outcome. Congress produced a deeply flawed piece of legislation. EPA failed to implement the new law vigorously. RCRA's rather inauspicious beginnings would prove to have serious repercussions later, once hazardous waste became a passionate mass issue.

Not Yet a Well-Formed Issue

At the time Congress acted, little was known either about the amount of wastes industry was generating or about those wastes' potential impacts on people's health. Public opinion had not yet developed. No organized forces campaigned to have Congress regulate toxic industrial waste.

A Dearth of Information

Before RCRA, no law required firms to report their waste practices to anyone, and they did not do so voluntarily. As the editor of the journal *Hazardous Waste and Hazardous Materials* remarked later, "Manufacturing industry that generates hazardous waste, in particular the chemical industry, for many years kept the lowest possible profile on the issue. Company spokesmen never mentioned hazardous waste and no public reports were made" (1985c:viii). Consequently, as of the mid-1970s, no one knew exactly how much waste was being generated annually by American industry, what happened to those wastes (where and how they were being disposed), or how they might be affecting the environment or public health.

In 1974, the EPA attempted for the first time to estimate the total size of the waste stream. It reported that 10 million tons of hazardous wastes were being generated annually, an amount it said was *"larger than originally anticipated"* (EPA, 1974:1; emphasis added). In 1976, the EPA raised its estimate to 37 million tons; in 1979, to 56 million metric tons. Four studies published between 1983 and 1986 produced estimates ranging from 247 to 275 million metric tons. Experts from the U.S. Congress's Office of Technology Assessment (OTA) told a congressional hearing in 1987 that the figure could be as high as 1,000 million metric tons. In 1989, the American Chemical Society put the total somewhere between 580 million and 2.9 billion tons.[3] In light of these later estimates, it is clear that, at the time Congress acted, little was known about the generation of industrial wastes and that policymakers were given unrealistic information about the magnitude of the problem they wished to address.

Equally little was known about what happened to those wastes once they were produced. How were these materials disposed of? Where did they end up? Responding to congressional inquiries, the EPA tried to sum up what was known, or thought to be known, about disposal practices. It reported that a modest hazardous waste treatment and disposal industry had developed by 1974, but only about 24 percent of its capacity was being utilized because "adequate treatment and disposal . . . costs 10 to 40 times more than the environmentally offensive alternatives" (1974:x, 12). The EPA thought that the vast majority of wastes were being disposed of improperly:

> By far, the most common disposal practice for industrial wastes is disposal into or on the land. . . . piled or dumped on the surface, ponded, lagooned, landfilled or spread on the land. Injection into deep wells is also practiced. . . . there are countless instances of dumping of hazardous waste in sewers, streams, swamps, quarries, abandoned farms, etc. . . . The most common off-site disposal practice is to dump hazardous waste in a municipal land disposal site where it is mixed with municipal solid waste. . . . On-site disposal is frequently

characterized by simply filling large lagoons with semi-liquid wastes or by the stockpiling of solid (dry) wastes. (House of Representatives, 1975:775)

The U.S. General Accounting Office's assessment was equally bleak. It was generally thought that about 80 percent of all wastes were being treated, stored, or disposed "on site," that is, on the grounds of the generating firms themselves; the rest was shipped off site to commercial disposal or treatment facilities. GAO's study of off-site facilities found that "many sites are located on land that is considered to have little or no value for other uses, such as marshes, and sand and gravel pits, and it is such siting which poses the greatest potential for environmental damage" (1978b:1). About the remainder, the 80 percent of wastes disposed on site, GAO reported that "virtually nothing is known about industrial [on-site] waste land disposal sites" (1979:ii).

The House Commerce Committee described the information vacuum in blunt terms:

> To date there has been no survey or other wide ranging investigation of the sources of hazardous or potentially hazardous waste generation or disposal. As a result, little is known about the actual volume of hazardous waste being generated, the geographical distribution of the generators or the extent to which hazardous wastes are transported. Neither does the Committee or the EPA know where much of the waste which is clearly hazardous is being disposed of. (House of Representatives, 1976:6264)

Finally, hardly anything was known about the impacts of industrial wastes on public health. The EPA's report to Congress listed a number of instances of local groundwater contamination and cases of health problems—lung damage, cancers, birth defects—thought to be caused by toxic exposure. The House Commerce Committee's report listed about sixty such cases from twenty states. Although these compilations were suggestive, they were no substitute for comprehensive epidemiological studies. But, as there were no reliable data on generation and disposal, hence on *exposure*, systematic studies of health effects had not been and could not be done.

No Well-Formed Public Opinion

Hazardous waste was a local issue, here and there, in the years before the nation was made aware of the plight of communities such as Love Canal, New York, and Times Beach, Missouri. In some places, people had begun to suspect that improperly disposed wastes had contaminated their communities. In other communities, people were beginning to oppose the siting of new waste disposal facilities near their homes. But these were, at the time, isolated events. "Toxic waste" had not yet coalesced into a distinct entity in public opinion. Perhaps the best way to highlight what had *not yet* happened in public opinion is to contrast two surveys, one done in 1973 and the other in 1980, of public attitudes toward the classic risk issue, "How close is close enough?"

In 1973, the EPA sought to find out citizens' attitudes toward having a national disposal site (NDS) near them, that is, toward the idea of having a federally owned hazardous waste disposal facility in their community. Judged from the hindsight of

Table 1. Acceptable proximity of a national disposal site (in %)

	Willing	Unsure	Unwilling	Don't know
Within 1 mile	36.8	27.9	35.0	0.3
Within 5 miles	57.9	21.4	20.4	0.3
Within 10 miles	67.4	18.8	13.5	0.3
Within 25 miles	70.1	17.4	12.2	0.3

Source: U.S. Environmental Protection Agency, 1973.

knowing popular sentiment today, citizens back in 1973 appear to have been remarkably unruffled by the prospect of having a hazardous waste disposal facility for a neighbor: 60 percent of respondents favored or strongly favored placement of an NDS facility in their own county; 58 percent thought that such siting would either leave property values unchanged or actually increase those values. Table 1 shows that almost 60 percent of the sample were willing to live within five miles of a hazardous waste disposal facility. It seems clear that the term *hazardous waste*, had not yet entered popular discourse, had not yet been clearly defined in the popular imagination, had yet to take on the connotations it has today.

Compare the numbers in the table to the findings of a survey done in 1980 for the Council on Environmental Quality, displayed in Figure 1. The proportion of people willing to accept a disposal site within five miles of their home had dropped below 20 percent. The 60 percent acceptance mark is passed only at the distance of one hundred miles.

By 1980, attitudes toward hazardous waste disposal sites had come to resemble closely post-Three Mile Island attitudes toward nuclear power, the technology that risk perception researchers always single out as *the* exemplar of technology that inspires extraordinary levels of popular dread.[4] In contrast, if one drew the earlier survey results on Figure 1, one would see that back in 1973, the American public rated its willingness to live near hazardous waste facilities somewhere between its willingness to live near an ordinary industrial plant and a ten-story office building. This shift in people's perceptions occurred, I argue in the next chapter, in 1978, when network television began its coverage of the plight of the Love Canal community.

Little or No Organizing

Some grass-roots hazardous waste protests were happening already, but they had not yet gone beyond sporadic, isolated instances of strictly local protest. The kind of self-conscious, focused national movement that now exists would not develop until much later. The more traditional, mainstream environmental organizations were focused on resource conservation and recycling issues, such as the nonreturnable glass bottle and the aluminum can, on what would now be called *postconsumer* waste; they had little to say, at the time, about the potential environmental impacts of *producer* waste, or what is now commonly referred to as *hazardous, industrial, toxic* waste.

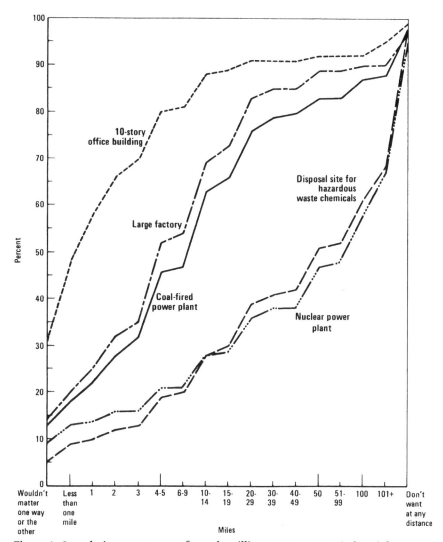

Figure 1. Cumulative percentage of people willing to accept new industrial installations at various distances from their homes (*Source:* U.S. Council on Environmental Quality 1980:31).

The Resource Conservation and Recovery Act of 1976

Onto the Legislative Agenda

If there was no public opinion and no social movement, if there was *no issue*, yet, how did Congress come to regulate hazardous waste? The step seems to have been largely a by-product of larger developments in national politics, a kind of legislative "free rider."

The modern environmental movement had generated considerable political momentum for expanding the regulatory apparatus of the state. In a mere handful of years, the federal government created agencies to protect workers (Mine Enforcement and Safety Administration, Occupational Safety and Health Administration) and consumers (National Highway Traffic Safety Administration, Consumer Products Safety Commission), and to monitor air quality, water quality, and other environmental conditions that affect the general population (Environmental Protection Agency) — eloquent testimony to politicians' perceptions of the salience and volatility of environmental issues at that moment.

Concern about potential waste and resource problems arose, first, within this larger context of increasing environmental awareness. Environmentalists criticized what they called society's throwaway ethic and warned that single-use, nondegradable packaging would deplete resources and soon choke municipal landfills. Local officials, too, worried about the landfill capacity problem. Some farsighted business and government officials were becoming concerned about the future price and supply of important raw materials. There was, then, growing support for addressing what were understood as "resource conservation" and "resource recovery" issues.

In response, Congress passed the Resource Recovery Act of 1970.[5] The act was intended, primarily, to involve the federal government in improving solid waste management. Secondarily, the act also expressed the beginnings of official interest in shifting society from its traditional reliance on disposal toward a new emphasis on the recovery and reuse of materials and energy. Almost as an aside, the act also requested a report on "the storage and disposal of hazardous wastes, including radioactive, toxic chemical, biological, and other wastes which may endanger public health or welfare" (*U.S. Code Congressional and Administrative News*, 1970:1434). The insertion of this phrase suggests that there were at least some in Congress who were beginning to see that there might be another kind of waste problem facing the nation. Still, it is clear that, at the time, waste legislation was primarily concerned with other, barely related, matters.

The Nixon administration opposed federal involvement in solid waste management and did its best to undermine the intent of the 1970 act.[6] Congress soon took up waste legislation again and the administration thought it could deflect attention from solid waste by shifting the discussion to another issue. The administration latched onto that peripheral paragraph of the 1970 act that called for a hazardous waste disposal site study. It would fight congressional liberals' renewed efforts to enlarge Washington's role in solid waste management; at the same time, it would argue that there *was* a legitimate need for federal regulation of a quantitatively much smaller but more dangerous waste stream, toxic industrial wastes.

Congressional proponents of waste and resource legislation had no problem with the latter half of the administration's position. It was not what they cared about most, but several instances of contamination caused by improper disposal of industrial wastes had come to light; clearly, it was a problem worth addressing. So, long before the toxic waste issue achieved the ripeness or maturity that issues usually must reach before lawmakers do something about them, everyone — president and Congress, Republicans and Democrats—suddenly agreed that it would be good to regulate industrial waste.

Conflict over Fundamental Questions of Policy Design

The era of uncontrolled waste disposal was about to end, but what kind of regulatory regime would be created? That would depend on how two fundamental design choices would be resolved during congressional debate: (1) Regulate production or regulate disposal? Would society deal with waste by regulating *production*, thereby making industry reduce the amount of waste it generates? Or would it forbear from interfering in production and focus, instead, on regulating and improving the quality of disposal? (2) Who ought to build, own, and run waste treatment and disposal facilities? Regulation creates *demand* for environmentally sound disposal practices; it does not automatically create a *supply* of appropriate facilities. Would the federal government take an active role in providing the supply of treatment and disposal facilities, as Scandinavian and Western European governments have done? Or would it be essentially passive, wait for private sector entrepreneurs to see the business opportunity, step in, and build the vast network of facilities that would be needed to make regulation work?

Regulate Production or Regulate Disposal? This was perhaps *the* fundamental question of policy design: Should society deal with its waste problem at the source, reducing the generation of wastes by regulating production processes and products, or should it, instead, emphasize improved disposal? The administration agreed that "control of toxic materials before they become toxic wastes could greatly reduce the size of the overall hazardous waste management problem" (EPA, 1974:21), but took the position that the Toxic Substances Control Act, then pending before Congress, would reduce toxics at the source and would "dovetail neatly" (ibid.) with a disposal regulation program. It proposed that generators' responsibilities be limited to packaging and labeling their wastes properly, keeping good records, and purchasing disposal services only from properly licensed hazardous waste facilities.

Democrats accepted the administration's "regulate disposal" approach to production process wastes; however, they proposed various forms of regulatory control over consumer products that would reduce the amount of solid wastes generated by throwaway packaging.[7] Senator Muskie's bill, the main alternative to the administration bill in the Senate, for example, would have regulated both product contents and manufacturing processes in the cause of reducing the ultimate volume of postconsumer waste: "Standards may include minimum percentages . . . maximum permissible quantities of component materials and may prescribe methods of distribution . . . and prohibitions against the manufacture and sale of specific items" (Senate, 1974:35). Other Democratic bills contained even more radical language on product and process controls. Representative Tiernan's bill proposed that

> the Administrator shall promulgate . . . standards regulating the manufacture and distribution of certain products in commerce as he determines necessary to protect health or the environment against unreasonable burdens and risks

associated with the disposal of such products. . . . The Administrator or the Attorney General may file an action [for seizure] against any product which constitutes an imminent hazard . . . or any product which the Administrator finds is manufactured or distributed in violation of section 203. (House of Representatives, 1974b:66, 68)

Now, although the Democrats had not proposed the source reduction alternative for *producer* waste, their proposals for product controls had obvious precedent-setting implications. If the government could interfere with private economic decision making (what to produce, how to produce it) in order to lower the volume of non-hazardous materials such as glass and aluminum, how could it not soon find it legitimate and necessary to interfere in production in order to protect citizens from *toxic, hazardous* industrial chemicals? The debate over how best to deal with the growing volume of postconsumer waste became the vehicle through which the larger question of generator control would be fought out.

The corporate sector made its stand on the terrain of beer and pop, bottles and cans, adamant that the nation should not solve its waste problems through regulatory control of production.[8] The sectoral breadth and ideological unity of the corporate mobilization was truly impressive. As hearings began in 1974, both the makers of consumer products that rely heavily on cheap, light, disposable packaging and the makers of the raw materials used in such packaging came out in force to oppose product standards. Members of Congress heard from the soft drink and beer industries; the American Iron and Steel Institute; the Stone, Glass and Clay Coordinating Committee; the Solid Waste Council of the Paper Industry; and Reynolds Metals. Perhaps the statement from Reynolds can stand as representative:

We are . . . extremely concerned about sections of those bills providing for Federal regulation of packaging materials based upon solid waste or other environmental considerations. . . . we believe that the packaging industry can best serve this Nation . . . in a free market environment, enabling the producer of products to select the package that, all things considered, best conveys his product to market. (Senate, 1974, part 2:389)

When hearings continued, a year later, manufacturers' testimony against controls over consumer product packaging again took up the vast bulk of House hearings. The American Iron and Steel Institute, the big automakers, the National Association of Retail Grocers and others representing retail food chains, the Glass Container Manufacturers' Institute, the Glass Bottle Blowers' Association, the Stone, Glass and Clay Coordinating Committee, and, of course, the National Soft Drink Association all made appearances.[9]

The industry perspective was backed by labor union spokesmen who argued that product standards, bans, or deposits would threaten jobs. The spokesman for the Aluminum Workers International Union expressed alarm that legislation "could put the aluminum can out of business" (House of Representatives, 1975:630). The steelworkers' lobbyist opposed what he characterized as proposals to "ban the can," and stated that "the entire concept of source reduction, it seems to us, may be premature at this time" (ibid.:648). Another union representative stated the union position quite plainly:

> I object to any standardization of products or packaging if it is going to elim- inate jobs. . . . We do not want to accept in any manner the elimination of our good jobs at the expense of the litter and solid waste problems [sic], no mat- ter how it comes about, whether it be standardization of products or what. We object to that. There have to be other ways and means in which we can do this without knocking our people out of work. (ibid.:671)[10]

Powerful industrial sectors, notably the chemical, oil, and plastics industries, came forth to tell both Congress and the EPA that they opposed any talk of govern- ment interference with economic decision making:

> We object to the absolute blanket authority to control production processes and product composition. . . . Orderly and timely investment of capital and replacement of plants would be seriously impeded . . . [it would lock in] tech- nology to one given point in time . . . create artificial cost and supply/demand distortions. . . . We fear the ripple effects of arbitrary decision would have far reaching social and economic consequences. . . . free market economics should be the primary force for stimulating recovery or recycling of materials. . . . Authority to control production, composition, and distribution of prod- ucts . . . would be devastating to free enterprise commerce. (Dow, in House of Representatives, 1974b:291-292; Senate, 1974, part 3:1478)[11]

> We believe that the disposal of wastes ought to be regulated instead of regu- lating the nature and use of the product or the type of manufacturing process used. . . . greatest emphasis should be placed on establishing standards which assure that the ultimate disposal method is satisfactory. . . . it is un- reasonable in most instances to require the use of certain types of processes solely based on the waste generated. . . . product standards could have severe economic effects . . . not be in the overall interest of the consumer . . . a det- rimental effect on the development of new materials and innovative uses of existing materials. (DuPont, in Senate, 1974, part 2:454; EPA, 1976:72-73)

> Legislation should not impede the natural interaction of raw materials, mar- ket and other forces that ultimately control the nature, quality, price, and suc- cess of products developed in our free enterprise system. (Union Carbide, in Senate, 1974, part 3:1748)

> The generator should be free to decide whether to treat or dispose of wastes. (Manufacturing Chemists Association, in EPA, 1976:565)

> No specific requirements or prohibitions should be set governing the recovery, reuse or disposal of industrial wastes. . . . Generators should be free to increase or decrease waste production rates, terminate waste production, treat their own wastes, and negotiate treatment or disposal service contracts in a free and com- petitive market. (American Petroleum Institute, in ibid.:1406, 1410)

In effect, EPA would be given power to prohibit the introduction of certain products into commerce. . . . attempting to control disposal by setting standards for the product would, we believe, prove to be expensive, cumbersome and unworkable. Such intervention in the marketplace would, inevitably, work to the detriment of the economy and the consumer. (Society for the Plastics Industry, in House of Representatives, 1974b:316)[12]

All this was in response to proposals dealing with product standards and solid waste. No one had proposed anything analogous for "hazardous" industrial wastes, that is, *process* standards that would reduce producer wastes. Some Democratic bills did propose requiring *permits* for hazardous waste generation. Permits would give administrative agencies some control over production; thus permits, too, had to be opposed.

The permit system for disposal facilities for hazardous waste seems appropriate. However, we strongly maintain that a permit system for generators of waste is unneeded. A regulatory system for generators of waste would unduly restrict American capacity to respond to needed changes by tending to "lock in" processes according to the technology available at the time the permit was issued. . . . The regulatory program should concentrate on standards for the actual disposal of wastes. . . . regulation of manufacturing processes must be avoided. (Dow, in ibid.:290-291)

We feel that permits should only be required of the disposal site operator. (B. F. Goodrich, in Senate, 1974, part 3:1441)

We support the provision for permits for operators of a hazardous waste disposal site. However, we consider permits for the generation of hazardous wastes to be unneeded, and could result in unnecessary restriction of manufacturing operations. (Union Carbide, in Senate, 1974, part 2:464)

Generators were not only adamant about wanting no regulatory control of production, they made it clear that they did not want to be held legally liable for what ultimately happened to their wastes. They were willing, they said, to package and label their wastes and to hire only disposal firms that were properly permitted by authorities. Liability, they argued, should then pass to the party in physical possession of the wastes. Consider the following statements:

We agree that the generator has some responsibility in the area, . . . [that is, to] make some determination that the disposer is competent and has the proper permits for disposal. . . . However, the waste hauler and disposer have the responsibility to assure, respectively, that the wastes are delivered for disposal at the proper location and are properly disposed. Irresponsible action is invited if the person holding the waste has no responsibility for it. (DuPont, in EPA, 1976:73-74)

[The generator should] confirm the competence and reliability of transporters, treaters and processors to whom the waste may be transferred. . . . Each transporter, treater and disposer should be responsible for his individual activities while the waste is in his possession. (Monsanto, in ibid.:410-411)

MCA recommends that the responsibility for the waste should be associated with physical possession of the waste, so that the generator should not be held liable for negligence of the transporter and the disposer of the waste. (Manufacturing Chemists Association, in ibid.: 565)[13]

Generator control proposals could not survive in the face of such corporate mobilization. Congress tends to avoid intruding into production if there is the slightest possibility of achieving legislative goals without doing so; certainly, members of Congress seldom commit the faux pas of continuing to pursue an intervention that is so clearly disliked by powerful elements in the private sector. If RCRA was to have any chance of passage, proposals to solve the nation's waste problems through generator controls had to be dropped. Kovacs and Klucsik, two House committee staffers, wrote: "Since the nonreturnable beverage container issue was such a volatile one, both the proponents and opponents of nonreturnable beverage container legislation agreed not to offer their proposals as amendments to H.R.14496. It was fortunate they reached that agreement for *no issue could more certainly have jeopardized the passage of H.R.14496*" (1977:259; emphasis added).

The Senate committee report rejected regulation of production explicitly:

The Committee recognizes that there are many unanswered questions about alternatives to reduce the volume of solid waste. Some alternatives look toward the initial phases of the manufacturing process—to materials used and product design. Attempts to regulate industrial and commercial operations have broad implications. The bill does not establish any Federal regulatory authority with respect to decisions in the manufacturing process. (Senate, 1976a:5)

On the floor of the Senate, Senator Randolph, one of the bill's main sponsors, reassured his colleagues that his committee "deliberately avoided the issues raised by proposals which would prohibit the use of certain types of containers or require deposits" (Senate, 1976b:21402), and Senator Stafford, the bill's other main backer, affirmed that he always wanted a bill that would establish

an absolutely even-handed Federal program [that would not] show a preference for one or another philosophical approach to resource use and disposal. . . . I am pleased that those who mistakenly perceived my position as one that would penalize certain products ["industry and union representatives who originally strenuously opposed my amendment"] have come to understand that was never the case. (ibid.:21403)

The House Commerce Committee report rejected source reduction for both postconsumer and producer wastes. In the case of the latter, the committee could not have been clearer:

Rather than place restrictions on the generation of hazardous waste, which in many instances would amount to interference with the productive process itself, the Committee has limited the responsibility of the generator for hazardous waste to one of providing information. . . . there will be no requirement of the generator to modify the production process to reduce or eliminate the volume of hazardous waste." (House of Representatives, 1976:6264-6265)

Who Would Provide Facilities? If the disposal regulation approach were to have any hope of working, many, many new facilities—sanitary landfills, chemical landfills, incinerators, waste treatment facilities, recycling plants—had to be built. Would the government itself provide the waste treatment and disposal capacity needed? Would it site, construct, own, and operate facilities? Would it play a more restricted but still active role, provide federal land for sites, facilitate siting through preemption of local land-use laws and eminent domain, assist construction with subsidies, grants, and loan guarantees? Or would it limit its role to setting and enforcing standards and perhaps providing some technical assistance to state and local governments and private entrepreneurs?

Back in 1970, when Congress asked the EPA to evaluate the concept of national disposal sites, lawmakers seemed to entertain the idea that the federal government might own and operate a system of waste facilities. That it should do so was not an inherently chimerical notion. The U.S. government had, in times past, built and operated bold infrastructural projects. Scandinavian and Western European countries had built superior waste management systems around a core of government-owned and -operated facilities.[14] Realistically, however, Congress had become increasingly reluctant to get involved in, or even financially support, infrastructural developments, especially if they involved unpopular, potentially controversial facilities.[15]

The fight over funding for solid waste and resource recovery facilities shows pretty clearly how little Congress was likely to do along these lines. Democrats wanted to provide grants and loan guarantees to help build solid waste disposal systems. They also sought $2.5 billion in loan guarantees to help build waste recovery plants. The administration and Republicans in Congress were, however, so adamantly opposed to any significant commitment of federal monies for facility construction that the Democrats had to begin to make concessions just to get the bill out of committee, and they caved in more and more as the bill moved toward passage. By the time House and Senate conferees finished negotiating, federal funding for solid waste and resource recovery plants had been slashed to a mere $35 million to build a few demonstration plants.[16]

Given that history, it was hardly likely that Congress would seriously consider federal ownership and operation of a national network of hazardous waste disposal sites. The administration position was that government ownership *would* work, and had to be rejected precisely because it would: "Public funding can obviously create the necessary treatment and disposal facilities by outright grants and other means . . . [but that would bring the] public sector . . . into competition with the existing hazardous waste treatment and disposal industry to the detriment of the latter and, if the approach is successfully used, will eliminate that industry" (House of Representatives, 1974a:78-79). If government would stay out of the field, however, the private sector would see the opportunities provided by regulation-driven demand. It would rise to the occasion and provide the nation with adequate disposal capacity.

Government should do as little as possible: "No Government action to limit the uncertainties in the private sector response is appropriate at this time" (EPA, 1974:x). The administration's position on this was never seriously challenged. Even much more modest proposals, such as giving EPA the legal right to secure sites through eminent domain, were only "casually debated" in Congress before they were "summarily rejected" (Walter and Getz, 1986:235).

The Final Design: Regulating Hazardous Waste "from Cradle to Grave"

By 1976, both major design issues had been resolved. The federal government would regulate disposal, not production; it would create a demand for, but would not take an active role in providing, the network of facilities that would be essential to make the new system work. The act was passed by overwhelming majorities in both houses and signed into law by Gerald Ford on October 21, 1976.[17]

It would be years before people understood the trouble those design choices would cause. At the time, RCRA's regulatory design seemed an appealingly straightforward and reasonable way to assert regulatory control over and assure safe disposal of hazardous industrial wastes. The law directed the EPA to identify and list substances that were to be considered "hazardous waste" and thus regulated (Section 3001). Section 3002 mandated the development of a manifest system, paperwork that would make it possible to document the movement of hazardous waste shipments "from cradle to grave"—from point of generation, through various transportation and temporary storage sites, to final disposition. Section 3002 also instructed the EPA to set standards for generators. Following Congress's intent to limit generators' responsibilities, the EPA interpreted this to mean that generators who shipped hazardous wastes off site would be required to identify the nature of their wastes (their composition and quantity), to package and label them properly, to start manifest paperwork for them, and to hand over their wastes only to transporters and disposal firms that had proper permits. Sections 3003 and 3004, respectively, authorized the EPA to set standards for transporters and for treatment, storage, and disposal (TSD) facilities. The EPA was empowered to enforce these standards by establishing a permitting process for TSD facilities.

Implementation

New regulations almost always address conditions that had been, until shortly before, by and large ignored. The infrastructure necessary for implementing regulatory controls—data, enforcement personnel, administrative organization, and so on—either doesn't exist or is grossly inadequate. In this respect, at least, the case of RCRA was entirely typical. Implementing the new law would have been a complex, difficult task in any case; lack of reliable information, numerous technical, scientific, and economic uncertainties, and the absence of any real infrastructure made things far worse.[18] However, the problems of early RCRA implementation cannot be attributed to inevitable start-up problems alone. The "prematurity" of legislative action was also a factor. Hazardous waste would not become a significant issue for at least two more years. Consequently, neither Congress nor the new Democratic administration made it a priority.

During the first oversight hearings devoted to the new law, in 1977, members of Congress seemed uninterested in RCRA's hazardous waste provisions. They did not actively question EPA about Subtitle C implementation until 1978, after a spectacular and newsworthy explosion occurred at Rollins Environmental Services in Logan Township, New Jersey.[19]

The Carter administration, for its part, was preoccupied with a worsening economy. Influential members of the administration, especially in the Office of Management and Budget (OMB) and the Council of Economic Advisers (CEA), argued that zealous pursuit of regulatory objectives was inflationary.[20] Because hazardous waste was not yet a vital issue, the administration sacrificed timely and effective implementation whenever that appeared to conflict with its economic objectives.[21] It continued to do so until events at Love Canal radically reordered its priorities.

In the name of budgetary restraint, the administration asked too little for RCRA implementation, then failed to use all of the already modest amounts authorized by Congress.[22] The GAO reported that

> EPA's hazardous waste program budget request for fiscal year 1978 included $14,450,000 and 195 staff positions for the development of hazardous waste guidelines and regulations and to implement the regulatory strategy. The Office of Management and Budget however, approved only $5,068,000 and 48 staff positions for the program. In a November 1976 letter to the President, the EPA Administrator expressed concern that the funding cut would prevent EPA from meeting deadlines established by the act. (1979:14-15)

As EPA staff wrote draft language that would be needed to implement Subtitle C (what wastes ought to be regulated; standards for generators, transporters, and disposal facilities), the rule-making process slowed because the proposed regulations were "expensive . . . much more expensive than the EPA administrators were prepared to defend" (Epstein, Brown, and Pope, 1982:226). Raising the price of disposal was exactly what the authors of the act had intended. After all, the EPA had pointed out, back in 1974, that the most common, and most "environmentally offensive" (1974:12), disposal methods are ten to forty times cheaper than better, environmentally sound alternatives. Now, Carter administration officials felt that rule making that raised the cost of waste disposal too steeply was unacceptable because it was inflationary. "John Lehman, head of the Hazardous Waste Division, was instructed by his superiors to cut back on the scope of . . . proposed RCRA regulations. Lehman ascribed the decision to the President's desire to fight inflation" (Epstein et al., 1982:228).[23]

Because draft regulations were withdrawn and extensively revised, the EPA missed every statutory target date set by Congress for issuing basic Subtitle C standards. Without these rules, the program could not begin to be implemented. Generators did not know what wastes would be regulated. The EPA could not begin to permit and inspect transporters and TSD facilities. Congress had wanted states to step forward and run hazardous waste programs that would be as good as or better than the basic federal program. But, the House Commerce Committee pointed out, "EPA's failure to promulgate regulations in a timely fashion is delaying State programs, as States are reluctant to act for fear that they will have to act a second time later to comply with as yet unknown Federal standards" (Senate, 1983, vol. 2:53). Furthermore, the financial and technical assistance promised to the states was not forthcoming, and that, too, contributed to paralysis at the state level.[24] A 1979 survey of the twenty-six states that, together, accounted for about 80 percent of the hazardous waste generated in the United States showed that

> all but 2 [state programs] are in very early stages of development or are wholly inoperable. Most States have not carried out even some of the basic first requirements of a hazardous waste program. . . . None of the 26 States had fully identified waste generators within their State jurisdictions, . . . know the exact volume of hazardous waste generated in their States, . . . none had adequate enforcement programs, . . . none could adequately account for the disposition of these wastes. (GAO, 1979:5-6)[25]

Slow, troubled implementation inhibited private sector initiative as well. The disposal industry was reluctant, understandably, to invest in new facilities without first knowing the standard these facilities would eventually have to meet.[26] Besides, unless they could be sure that regulations would be fully enforced, investing in facility construction was a poor risk. In the words of the executive director of the National Solid Waste Management Association: "Without enforced regulations, there is little incentive for generators of hazardous waste to pay the high cost of disposal at specially designed and designated facilities. . . . Until a hazardous waste management program is in place and enforced, there will be minimal development of new facilities" (GAO, 1978a:4).

The political context changed in 1978, following media coverage of Love Canal and other similar chemical pollution disaster stories. The EPA came under intense scrutiny and criticism from Congress and the environmental community. The U.S. General Accounting Office, Congress's investigatory arm, produced a series of studies highly critical of EPA's efforts to implement RCRA's hazardous waste provisions.[27] Oversight hearings featured discontented EPA staff airing their grievances and members of Congress heatedly demanding that the administration do better in addressing an issue so important to the American public.[28] Carnes observed that "It is likely that, since 1978, Administrator Costle and his deputies and their staffs have spent more time on Capitol Hill testifying on Love Canal, EPA's hazardous-waste activities, RCRA regulations, and the lack of regulations, than on any other EPA mission" (1982:41).

Congress's case against RCRA implementation was summarized in the Commerce Committee's report on HR7020, a Superfund proposal. The committee charged that the EPA had failed to conduct a comprehensive inventory of the location and content of all places where hazardous waste has been disposed; . . . Although the Congressional deadline of 18 months for EPA promulgation of regulations may have been unduly short, there is no excuse for not having the regulations promulgated by now, three years later. . . . EPA has proposed inadequate regulations . . . [that] fail to list as hazardous a number of known carcinogens; . . . wrongfully exempt some generators of highly dangerous hazardous wastes simply because the quantity of wastes disposed is not large . . . [and] permit the granting of interim permits to sites that are far below the minimum safety level. (Senate, 1983; vol. 2:52-53)

The administration had to respond. As Steven Cohen and Marc Tipermas, two political scientists who were EPA staffers at the time, later wrote, "As the magnitude of the environmental problem became more obvious, and *the potential for political trouble became clearer*, EPA's top management began to see the cleanup of existing hazardous waste sites as a critical environmental problem" (1983:47-48; emphasis added).

The administration seized the initiative on the now-hot issue. It proclaimed hazardous waste one of the most serious environmental problems facing the nation. The EPA and the Justice Department jointly formed the Hazardous Waste Enforcement Task Force to locate potentially dangerous sites, evaluate them, and bring suit against those responsible. The administration took the lead in calling for a Superfund to deal with the problem of abandoned, leaking hazardous waste dumps.[29]

Finally, implementation of RCRA's hazardous waste provisions was being pursued with some vigor. However, in spite of some progress, every facet of the program continued to be problem-ridden. EPA issued its proposed regulations in December 1978. Final rule making was then further delayed when industry responded to the administration's draft regulations with more than 1,200 sets of comments, "some of which rivaled a New York City telephone book in length," an outpouring of comments "unprecedented in EPA history" (Lehman, 1984:12). EPA officials complained "privately" (that is, anonymously, to a reporter from the *New York Times*) about the "millions of pages of testimony filed by representatives of industry on virtually each clause of every implementation proposal" (Shabecoff, 1979:1).

EPA finally managed to issue some of the required RCRA standards, two years late, in February and May of 1980. Even then, it left out an absolutely central part of the rule making, technical standards for treatment and disposal facilities. The agency said that would take several more years; until then, all treatment and disposal facilities would have to meet only "interim" standards that were "largely administrative and nontechnical and were not intended to provide complete health and environmental protection" (GAO, 1981:ii).

Given that, it is perhaps unnecessary to criticize enforcement; even if relentless, enforcement of "largely administrative and nontechnical" standards cannot mean much. Still, a review of GAO's criticisms of the EPA's inspection and enforcement program will round out our picture of RCRA implementation as of about 1980. According to GAO, the EPA had no idea if all facilities required by law to apply for interim permits had done so. When facilities did apply, interim permits were granted routinely, without any critical evaluation of the application. Once permits were granted, few facilities were inspected:

> Only about 12 percent of the 7,056 facilities with interim status in the four regions GAO reviewed had been inspected by EPA and/or the States, and these inspections had been primarily of administrative provisions of the interim status requirements . . . [even though] analysis of 127 inspection reports in the four regions showed that 122 of the facilities inspected did not comply with the interim status regulations. (ibid.:iii-iv)

RCRA in Context: Analyzing the Roots of Regulatory Failure

For the past twenty years, the modern environmental movement has been fighting, often successfully, to expand the array of regulatory laws that are meant to protect the environment.[30] Absolute gains have been achieved in some cases. Clean water legislation has brought some lakes and rivers back from the brink of death. The haze blanketing the nation east of the Mississippi has eased somewhat.[31] Undoubtedly, too, there have been gains relative to the level of damage we would now be seeing had there been no regulation of industry at all during that twenty years.

If we examine *procedural* and *discursive*, as opposed to just *substantive*, impacts, our evaluation must be even more positive. Regulatory agencies generate invaluable data about the state of the environment. Further, the agencies provide a focus for media coverage and ongoing political scrutiny: Is the agency doing its work? Is it protecting us sufficiently? If not, what and how extensive are the problems that it is not dealing with adequately? This helps keep environmental issues at the center of national political discourse. Procedurally, regulatory laws provide new administrative and judicial opportunities for citizens who wish to contest commercial development or industrial activity. The Environmental Impact Statements mandated by NEPA, the National Environmental Protection Act, are just one prominent example of the kinds of procedural tools provided by such legislation.

Still, discourse and procedure, talk and litigation, are only talk and more talk unless they ultimately produce substantive outcomes. Barry Commoner recently asked "an important and perhaps embarrassing question: how far have we progressed toward the goal of restoring the quality of the environment?" His answer: "Apart from a few notable exceptions, environmental quality has improved only slightly, and in some cases worsened" (1989:12).

Why is it that twenty years of heightened regulatory activity have produced, at best, only slight improvement? We have at our disposal a century of historical experience with regulation. That experience has been studied extensively by historians, sociologists, political scientists, and economists. Social scientists' gaze has not been kind to regulation. It would hardly be an overstatement to say that the study of regulation has been coextensive with the search for the causes of regulatory failure.

"Protest and Regulate": The Logic of Environmental Crisis Management in Capitalist Society

The problems of regulation can be traced, first, to its historical origin as a corrective to the fundamental structural relationship between economy and society in capitalist societies. In earlier social formations, economic activity was embedded in, fused with, and constrained by other social/institutional arrangements.[32] Capitalism required the disarticulation of economic activity from traditional feudal restraints. Karl Polanyi described the core imperative of this new economic ethos in this way:

> Nothing must be allowed to inhibit the formation of markets. . . . no measure or policy must be countenanced that would influence the action of these markets. Neither price, nor supply, nor demand must be fixed or regulated; only such policies and measures are in order which help . . . [create] conditions which make the market the only organizing power in the economic sphere. (1944:69)[33]

Theorists of the market argued that external restraint on production in the name of collective, societal interests and conditions was unnecessary and undesirable. Individual economic actors had to be free to choose what to produce, how much and how, guided solely by their desire to maximize private gain. The laws of political economy would guarantee, they said, that the sum of individual, self-interested economic activities would produce, even optimize, collective, societal good.

Well, in fact, experience showed, from the earliest days, that the invisible hand could not be trusted to author an unequivocally benign text. The new economic arrangements demonstrated a fantastic capacity to generate wealth; they soon began, however, to show a predilection for overconsuming and abusing natural resources. In words all the more remarkable because they were written half a century ago, Polanyi argued that such results were necessary and inevitable:

> Machine production in a commercial society involves, in effect, no less a transformation than that of the natural and human substance of society into commodities. . . . [But] Labor is only another name for a human activity . . . land is only another name for nature [p. 72]. The dislocation caused by such devices must disjoint man's relationships and threaten his natural habitat with annihilation [p. 42]. Robbed of the protective covering of cultural institutions, human beings would perish from the effects of social exposure; they would die as the victims of acute social dislocation. . . . Nature would be reduced to its elements, neighborhoods and landscapes defined, rivers polluted . . . raw materials destroyed [p. 73]. (1944)

When growth in the manufacture of woolens stimulated the market for wool, for example, common lands were enclosed and transformed into sheep runs that turned "overburdened soil into dust" (ibid.:35). Extraordinary extensions of the working day, factory machines without safety guards, cramped, hot, unventilated, unhygienic shop-floor conditions had appalling impacts on the health of the laboring classes.[34]

The environmental and public health impacts of unfettered economic activity could not be ignored. Something had to be done, but what, and how much? Economic activity—production and markets—had just been jubilantly liberated from traditional constraints. Laissez-faire ideologies were ascendant, hegemonic. Could anyone have reached the idea that the newly established relationship between collective and private interest was fundamentally flawed and that economic action must be reinserted in the web of other, constraining, institutional arrangements? The most that proved possible was a *partial* reassertion of social/political control through the imposition of state "regulation," and even that was done with "hesitation [and] repugnance" (Marx, 1867/1967:494).

It began with hours legislation. Factory owners had increased the length of the working day. Factory hands were being driven to the point of exhaustion. Without adequate opportunity for rest and sleep, they were unable to recover from the exertions of one day before the start of the next. The health of the working people of England rapidly deteriorated. Enlightened observers worried about the capacity of a nation to survive such an attack on its human substrate; even those not so far-sighted could see that these practices were threatening to unleash social unrest. Parliament stepped in to regulate the hours of work, first for women and children, then for all workers in specific industries. Subsequently, when new forms of damage were discovered and people demanded that something be done, factory acts and shorter hours legislation served as the model.[35] "Protest and regulate" became the fundamental structure of environmental protection in capitalist society.

Regulation: the word says it all. The regulator added onto the steam engine does not fundamentally change the engine; it prevents catastrophe, at the last moment, by relieving pressure when it builds to the breaking point and threatens to blow up the whole works. Regulation did not fundamentally modify the presumption of laissez-faire; it merely added an error-activated safety valve. Private actors were still free to undertake economic activities without considering their potential adverse impacts on nature, collective conditions, or future generations. Raw materials were mined as cheaply as possible, untreated streams of effluent were discharged into air and water. Because nature has a large, if finite, capacity to absorb damage, hazards accumulated for decades without being recognized.

Nothing happened until nature's capacity to absorb the damage was overwhelmed. Only then did society begin to understand that something had happened to the air, to the water, to food supplies, to other species, to people's health. Only then did significant social forces mount protests and make political demands that government do something about the harm that had been done.

Nicolas Ashford, in a well-known study of contemporary worker safety and health regulation, posed this question: Should we prohibit use of a substance or technology until we know it is safe, or should we go ahead and use it until science is certain that it is harmful?[36] Naive common sense would suggest that, if we really could choose, we ought to choose the former. Given that we have only one viable planet within our reach, assuring the collective conditions of survival on earth ought to be societies' first and most dearly held goal. Preserving the necessary natural conditions of human existence ought to be *the* dominant consideration in any societal decision making that could possibly affect those conditions, perhaps along the lines of what is said to be the traditional Iroquois criterion, "Will this be to the benefit of the seventh generation?"

Ashford phrased his question as if the matter were still open, as if it were a question that "we" as a society would have to debate and resolve, as if the choice had not already been made. In fact, that choice was made long ago and is now woven into the very fabric of modern political economy. Therein lies the first, and perhaps most fundamental, problem of regulation. The logic of regulation is not a logic of prevention; it, in fact, embodies the polar opposite of that logic. Protection of nature depends on a political form that is inherently error activated and after the fact. Given that, regulation can hope to be effective only if all important forms of damage come to society's attention promptly, if awareness is quickly followed by regulatory response, if that regulatory response is strong enough to be effective, and if the law is then fully implemented. Studies of regulation suggest that none of these conditions is routinely met.[37]

Difficulty and Delay in Achieving Societal Realization. Before anything else, the mute physical fact of damage must somehow be made an element of political discourse. Someone, be it the people directly suffering from the situation, concerned social observers, or, perhaps, scientists, must see a pattern of effects and begin to suspect a common cause. They then have to convince many others that the damage is real, that it is as grave as they claim, that it is caused by this or that economic activity.[38] The issue or problem must then be cast in political terms, organized as a demand for new regulatory action by the state. If all of these things happen, change may move forward; however, there are serious impediments at every step.

Problems can go unrecognized for long periods. They can be masked by nature's capacity to absorb damage. The sulfur dioxide emitted when coal is burned to produce electrical energy fell as acidic precipitation for decades, until the soil's absorbing capacity was exceeded and overwhelmed; at that point, lakes and forests began to die and science discovered "acid rain." Damage from industrial activity can go on for very long periods and become quite extensive if it happens far from any immediate human experience. Chlorofluorocarbons accumulated in the upper atmosphere for decades and had already begun to deplete the earth's ozone shield before scientists could gather the data and persuade society that a disaster was in the making.

Even when the damage is immediate, visible, and personal, causal attribution can be difficult. It took decades to confirm, for example, that workers' illnesses were being caused by exposure to chemicals in the workplace.[39] People lived near chemical dumps for years without attributing their health problems, ranging from rashes to miscarriage and birth defects, to the proximity of toxic wastes.[40]

When a first circle of people do become convinced that a problem exists, they then have to persuade a significant segment of society that the risk is real and serious, and that it is really caused by the industrial activity identified. The firms or industrial sectors blamed for the problem are likely to respond with counterclaims — that there is no problem, or that it is exaggerated, or that the data are too poor for anyone to say anything certain about the purported risk.

Advocates of regulation find it difficult to win these definitional struggles. Documenting the fact of damage, estimating the dimensions of the risk to environment and public health, establishing the causal connection between some industrial activity and the damage, and estimating the benefits and costs of regulatory intervention are all issues fraught with scientific uncertainties. How many cancers does workplace exposure to certain chemicals cause? Are the health problems of people living near chemical waste dumps actually caused by these dumps? Is the greenhouse effect already upon us, or is the warming trend just normal fluctuation and the greenhouse an unproven conjecture? Almost every claim that a risk is present, almost every attribution of cause, can be and often is vigorously contested.[41]

Even if this battle for definition is won, the formation of policy is still far from guaranteed. The firms or industrial sectors facing the prospect of being regulated argue that regulation will raise prices, make it impossible to continue production, put people out of work. They may be powerful enough to delay the move from issue creation to legislative action, typically for years, sometimes much longer.[42]

To summarize: the first problem with the logic of "protest and regulate" is that getting to the point of regulation is itself an arduous, and by no means certain, process. Nothing guarantees that all damage worthy of attention will come to society's attention in a timely fashion. There are great difficulties in moving from the fact of damage to politically salient awareness of damage. Even when the process gets to the point where demands for legislation are organized, powerful pressures can be brought to bear to try to get the state to abstain from legislating.

Limits Imposed in the Act of Legislating. The difficulties just described were more serious in the past than they are today. Until recently, almost all forms of damage languished in the twilight state of physical fact but social/discursive nonexistence. Modern environmentalism has moved society decisively beyond that point. It has succeeded in creating a generalized belief that economic activity typically produces at least some environmental harm. That presumption, in turn, makes it easier, even routine, to raise specific claims and demands for new government action. As a result, many new regulatory laws have been passed. The fact of legislative action may mean little, however, if the interventions are limited, too meek to deal seriously with the problems at hand.

Structuralist analyses of the state in modern society assert that fundamental limits on the extent of state intervention are imposed at the very moment the intervention occurs.[43] The key is government's dependence on the accumulation process and, hence, on business confidence. Claus Offe argues that

> in the absence of accumulation, everything, and especially the power of the state, tends to disintegrate. If we think of the budgetary obligations of the state . . . , its extensive reliance on resources created in the accumulation process and derived through taxation from wages and profits, this becomes immediately clear. Thus, every interest the state (or the personnel of the state apparatus, its various branches and agencies) may have in their own stability and development can only be pursued if it is in accordance with the imperative of maintaining accumulation. (1975:126)

Offe says that this "dependency upon accumulation functions as a selective principle upon state policies" (ibid.).[44]

Fred Block argues that this connection between state dependence on accumulation and the avoidance of certain policies is concretely achieved through state managers' sensitivity to fluctuations in "business confidence" (1984:38).

> Business confidence . . . generally responds unfavorably to an expansion of the government's role . . . in regulating the market [p. 42]. A sharp decline in business confidence leads to a parallel economic downturn . . . High rates of unemployment . . . annoying shortages of critical commodities. The popularity of the regime falls precipitously [p. 39]. If the state managers decide to respond to [strong popular] pressure with concessions, they are likely to shape their concessions in a manner that will least offend business confidence [p. 42]. (ibid.)

Concern about business reaction to the imposition of regulation is not idle. At the level of the individual businessperson, research shows that regulation violates deeply held beliefs in laissez-faire and in the sanctity of private property.[45] The businessperson finds regulation "frustrating" because it constrains "ordinary commercial and industrial dealings" (Lane, 1966:30). More globally, regulation is experienced as an attack upon the businessperson's ego that stimulates reactions of "anxiety, frustration, dejection beyond all relation to the economic costs" (ibid.:19). At the level of the firm, proposals for regulation inevitably bring forth warnings that there will be dire consequences—higher prices, closed factories, lost jobs—if the proposed regulations seem too onerous.

The structural situation of the state expresses itself to lawmakers as the felt imperative to strike a balance, to try to do something that will satisfy the demand for regulation but not impose unacceptable levels of cost and constraint on economic activity. This means, in the first place, that interventions that would fundamentally redefine the economy/society relationship by reasserting "social . . . governance of the means of production" (Commoner, 1989:12) are excluded almost automatically.[46] In practice, this means that instead of stopping pollution at the source, policymakers will choose to try to control pollution at the end of the real or figurative tailpipe, a regulatory strategy that, most authorities now agree, does not work.

Further, once the basic design issue is favorably resolved—that is, pollution reduction is once again rejected in favor of pollution control—firms then fight to weaken the bill's provisions, to secure statutory language that will minimize their costs and minimize the controls that will be placed upon them.[47]

Firms will not get everything they want, but vociferous opposition and dire predictions of adverse economic impacts secure further concessions. Debate may begin with strongly worded proposals, but members of Congress will get up and say, "Yes, we want to protect the environment, but none of us really wants to drive companies overseas or hurt national economic performance or make it impossible for corporations to function." The bill's sponsors must make compromises if they hope to get the bill out of committee and secure passage by both houses.

That does not mean that the regulation will be totally compromised and ineffective. It does mean that policy will try to deal with the problem in a way that minimizes government interference and minimizes the financial burden on the regulated firms. It does mean that substantive and procedural concessions are inscribed in every detail of the regulatory intervention.

Further Limits Imposed by Incomplete Implementation. Even if policymakers do find the will to make a relatively strong law, the theoretical magnitude of the intervention is realized only if the statute is fully and vigorously implemented. Social science studies of regulation suggest that implementation of environmental laws invariably falls far short of fulfilling legislative expectations. Broadly speaking, research on regulation finds two causes for the chronic problems of implementation. One type of explanation focuses on the inherent problems of bureaucratic organization and administration; the other emphasizes the impact of strategic resistance by regulated industries.

Political scientists who have studied government bureaucracy and administrative behavior argue that, even under the best of circumstances, administrative bureaucracies are inherently incapable of functioning effectively.[48] (The title of Lindblom's article, "The Science of 'Muddling Through,'" conveys well the thrust of such analyses.) New regulatory agencies rarely enjoy the best of circumstances. The complex, far-reaching problems that agencies are told to deal with have been, until that moment, neglected by society. Typically, there is an information vacuum. The overall scope of the problem is hardly known. The detailed scientific data needed to write sound standards barely exist. The agency has little to guide it in deciding what to do first, which standards to develop, where to send its probably small and insufficiently trained corps of inspectors.[49] Agencies spend years trying to generate the infrastructure necessary even to begin to do the job assigned to them.

Insufficient budgets, even under friendly presidential administrations, antagonistic political leadership during unfriendly ones, only exacerbate the problems. Suffering from the combined impact of organizational, situational, and fiscal problems, agencies cannot begin to cope with a mandate that requires them to monitor thousands of substances, billions of pounds of pollutants, entering the environment at thousands, in some cases millions, of widely dispersed sites. Administrators are constantly forced to engage in a kind of implementation triage, forced to make (typically poorly informed) choices between what to act on and what to walk away from. They never manage to carry out more than a small fraction of the work.[50]

Ongoing industrial or corporate resistance is the second reason offered in the literature to explain weak implementation. This strand of regulatory critique has a long pedigree that goes back to classic studies of the "life cycle" of regulatory commissions and to theories of regulatory agencies "captured" by the industries they are supposed to regulate.[51] Today, this line of analysis emphasizes two things: first, procedural, due-process safeguards are routinely inserted in the statutory language of regulatory laws;[52] second, regulated firms, especially large corporations, have the legal, technical, and administrative resources to use these procedural safeguards to slow down and weaken implementation.[53]

Vulnerability to Periodic Takebacks. Even the limited regulatory protection that can be achieved may not be secure. It has been observed that regulations come in "waves," clustered in relatively brief historical moments when political instability makes it prudent to make concessions to popular discontent. One such wave occurred earlier in this century, during the New Deal; a second, around 1970, was triggered by the general social unrest of the latter half of the 1960s.

Business interests never really reconcile themselves to the imposition of external control. As Miliband mordantly observes, concessions made during politically vulnerable moments are "part of that 'ransom' [which has] to be paid precisely for the purpose of *maintaining* the right of property in general. In insisting that the 'ransom' be paid, governments render property a major service, though the latter is seldom grateful for it" (1969:78). It can be expected, then, that when a favorable conjunctural moment arrives, when protest wanes and business interests are confident in their control of the political and ideological situation, regulated firms will agitate "for a restoration of the *status quo ante*" (Block, 1984:43) to roll back the clock, to get lawmakers to take back the concessions made during earlier periods of threat. Witness the deregulatory agitation that began around 1975 and reached fruition in the first Reagan administration.[54] When successful, such moments of takebacks undo much of what had been accomplished earlier.

RCRA as Another Case of Regulatory Failure

It should be clear that RCRA was a fairly typical case of regulatory failure. It deviated from type in only one way: events conspired to produce legislative action before hazardous waste became a mass issue and before political demand for regulation was mobilized. That might seem, at first, a significant departure from the expected pattern. In reality, it did not make much difference in the overall picture. Even if the legislative moment arrived years earlier than it might have otherwise, the reversal of the issue-creation and legislative moments merely ensured that limits would be all the more firmly imposed at subsequent points.

No issue, yet, meant that there was no well-formed public opinion, no movement, no aroused constituency in whose name legislators could fight. Even the established environmental lobbies had little to say about industrial waste management.[55] There were no forces that could put forth a plausible, politically viable alternative to the administration's policy proposal; hence its approach to regulation, strongly favored by corporate generators, was adopted without serious challenge, modification, or compromise.

Implementation would likely have been plagued with difficulties in any case, but, had the issue been as compelling as it would later become, there would have been motivation to do better. There could have been bigger budgets, more stringent standards, more extensive enforcement efforts. In actuality, because the issue had, as yet, no political presence, implementation was even slower and weaker than it had to be. RCRA could be starved of funds, sacrificed to other administration goals, without concern that there might be repercussions or political costs.

So it happened that a favorable deviation at one stage of the legislative process paradoxically made possible the imposition of failure-inducing limits at later stages. Later, years later, RCRA legislation would be a key element in a larger matrix of political events that would threaten to burst the limits historically inscribed in the "protest and regulate" form. How that happened, by what cunning of events and under what circumstances the routine reproduction of structural policy limits would begin to turn into its opposite and contribute to the transgression of those limits, is one of the core themes of subsequent chapters.

Repercussions

No one could foresee in 1976 that hazardous waste was about to become a spectacular mass issue. Before then, neither the policy design choices that were made nor the dreary first efforts to implement the new law had serious political consequences. However, when the words *hazardous waste* did become synonymous with extraordinary threat, both the flaws in RCRA's design and the failures of early implementation contributed to the further development of an increasingly stormy issue history.

The choice to regulate disposal rather than to encourage waste reduction made regulatory success dependent on construction of many new facilities. Demand for disposal and treatment sites could only increase as the economy grew, as new waste streams would be found hazardous, as RCRA set increasingly strict standards and old, noncomplying facilities shut down. The siting of enough new facilities would be *the* key to successful hazardous waste management.

Facility siting would depend, in turn, on public acceptance. If communities wished to resist siting, they would have considerable means at their disposal. Land use is traditionally under the jurisdiction of local government bodies that tend to be highly responsive to local sentiment; in addition, RCRA contained extensive public participation provisions. Siting could succeed only if local citizens' trust and consent were secured. What would happen if the public came to fear hazardous waste? Every siting attempt would become an ordeal.

Legislation had created a tinderbox situation where, if things went wrong, policy would stimulate opposition and opposition would threaten to destroy policy. By 1979, officials could see the implications all too clearly:

> Adequate treatment and disposal capacity is critical to carrying out the hazardous waste regulatory program. . . . public opposition to siting facilities is a major constraint to expanding disposal capacity and . . . it is likely to increase in the future. (GAO, 1978a:i-ii)

> Public opposition to . . . siting . . . is a critical problem. . . . [If it] cannot be solved . . . the national effort to regulate hazardous waste may collapse. . . . the prospects for successful sitings in most regions of the country are dubious at best, and grim at worst. (Centaur Associates study of public opposition, EPA, 1979:III, IV)[56]

Would federal ownership and operation of TSDs, or direct federal control of the siting process, have made a difference? In terms of the likelihood of public acceptance of facilities, probably not. Although the 1973 EPA public opinion poll showed positive attitudes toward the concept of living near "national disposal sites," a later study showed that people trust government regulators as little as they do private facility operators.[57] Recent protests about environmental hazards at federal nuclear weapons production facilities show that state ownership does not guarantee environmentally responsible operation; and they show that, once citizens living nearby find out they are at risk, it matters little whether the facility is publicly or privately owned.

The principal effect in this case, I believe, would be on the *form* of oppositional organizing. Centralized federal control of siting and facility operation would have tended to give rise to more centralized and formal forms of citizen participation. Decentralized siting, left largely to private initiative, favored the development, instead, of community-by-community confrontation between site developers and state regulators on the one hand and aroused local citizen groups on the other.

Decent implementation, and the resultant belief that government regulators were ensuring that facilities operate safely, could have eased the situation. But, as should be clear from the discussion above, RCRA implementation between 1976 and 1980 did nothing to allay citizens' concerns. Existing facilities were allowed to operate with interim licenses that had no substantive meaning; new, safer facilities were not being built. Officials' descriptions of the situation hardly instilled a sense of confidence: "We do not know where the millions of tons of stuff is going. . . . it is simply a wide open situation, like the wild west was in the 1870's, for toxic disposal" (James Moorman, assistant attorney general for land and natural resources, quoted in Worobec, 1980:636). The conjunction of the *anticipation* that stronger, more stringent regulation was just over the horizon and the *reality* of lax enforcement of interim regulations occasioned an outburst of what would come to be called "midnight dumping."[58] There were reports that organized crime was involved in hazardous waste hauling and disposal.[59] Media coverage of these problems eroded trust in regulators just when trust was most needed.

After Love Canal, after the proliferation of news stories about industrial waste, miscarriages, birth defects, cancers, perhaps nothing could have been done in the short run to calm people's concerns and fears. Opposition to siting was probably unavoidable. But poor implementation made things worse. It exacerbated all the negative possibilities inherent in a flawed regulatory design.

Chapter 3

"Toxic Waste" as Icon:
A New Mass Issue Is Born

The nation now had a law that, on paper, created a comprehensive system for the safe disposal of industrial wastes. However, exactly because hazardous waste had not yet become a significant social issue, government officials neglected that new law. Had nothing else intervened, we would most likely have witnessed an uneventful, entirely mundane story of continuing regulatory failure. At most, there would have been routine struggles over details, prodded sporadically, perhaps, by *60 Minutes* or a critical public television documentary.

Something else did happen. Localized protest concerning environmental threats had been occurring, here and there, for a decade. Until 1978 these actions had been strictly local affairs. Suddenly, one of these local actions broke into the national news. *Love Canal* and *toxic waste* became household words. Practically overnight, hazardous waste went from being a hazy, poorly organized perceptual object in popular imagination to being the most feared of environmental threats.

The Birth of a New Issue

Modern Environmentalism Altered Popular Beliefs

The first wave of environmentalism in the United States, dating back to the turn of the century, had focused on conservation, on sequestering and preserving extraordinary natural places. Unlike that earlier movement, the modern environmentalism that arose in the 1960s emphasized, instead, that industrial activity pollutes the "ordinary" environment everywhere and, in the process, threatens people's quality of life, even their very health. Books associated with that movement, such as Rachel Carson's landmark work on DDT, *Silent Spring*, and Ralph Nader's book about the hazards of the Chevrolet Corvair, *Unsafe at Any Speed*, showed in the most convincing fashion that the chemical miracles of modern production could have deadly environmental consequences and that even the largest, most legitimate corporations put profit before public well-being. Media coverage of dramatic pollution events, such as the Santa Barbara oil spill of 1969,[1] supported the movement's claim that unregulated industrial activity had ugly and disastrous consequences.

The modern environmental movement achieved considerable momentum by the end of the 1960s. It reached its symbolic pinnacle moment with the first Earth Day, on April 22, 1970. An estimated twenty million citizens took part in Earth Day activities. Media coverage of environmental issues reached new highs in 1970 as well. Polls indicated the impact on public opinion: By 1970, most Americans—on the order of 70 percent—agreed that air and water pollution were "very or somewhat serious" problems; a majority—53 to 55 percent—supported more government action to deal with pollution, even if that meant more spending and higher taxes.[2] Thus, years before Love Canal, modern environmentalism had produced a new set of beliefs about industry and pollution.

Throughout the next decade, the nation continued to hear sobering environmental news. Recall some of the most disturbing instances: the dioxin accident in Seveso, Italy; Kepone contamination in Hopewell, Virginia; PCB contamination of the Hudson River; the Binghamton, New York, office building fire that spread PCBs throughout the building; the accidental mixing of PBB-laden fire retardant into cattle feed in Michigan and the subsequent destruction of millions of head of cattle and the spread of PBB through the food chain in Michigan; and, of course, Three Mile Island. Television news also carried a steady stream of reports that many of the substances that people come into contact with, either as producers (chemicals, dusts, metals) or as consumers (notably, preservatives, pesticides, and other food residues), are toxic or carcinogenic.[3] Viewers were reminded again and again that unregulated economic activity can produce catastrophe, and that neither industry nor government could be trusted to keep the public safe.

Political Impacts? New Laws, Local Environmental Conflict

But what did these new developments in popular attitudes and beliefs really mean? Were the attitudes generated by modern environmentalism durable? Even if they were, did these attitudes have any real depth? That is, would they translate into actual political behavior?

Politicians certainly responded, initially, as if they believed they were dealing with something real and substantial. Congress quickly passed sweeping new laws to regulate air and water pollution, consumer products, worker safety and health. Even Richard Nixon sounded like an environmentalist.

Still, there were doubts about the depth and staying power of pro-environmental sentiment. Anthony Downs, an influential political scientist, argued that "issue attention" always takes a cyclical, up-and-down form in American politics. He predicted that interest in the environment would soon decline.[4] Others wondered if interest in the environment would survive in a time of growing economic turmoil. Modern environmentalism had achieved its successes, in part, because the nation's economy had been so vibrant for so many years; a widespread sense of security and well-being allowed noneconomic issues to rise in salience. But economic good times were about to come to an abrupt halt. The 1970s would bring energy crisis, recession, levels of inflation and unemployment not seen in decades. Polls soon showed that Americans thought that the state of the economy was by far the most important problem facing the nation; other polls showed that people were apprehensive about the future, that they expected that their personal fortunes would decline.[5]

Dunlap's reviews of polling data seemed to confirm fears that support for the environment would suffer.[6] Although environmental issues continued to have majority or plurality support during the 1970s, the degree of support decreased steadily as the decade wore on. Furthermore, Dunlap found that issue "salience" — that is, how much an issue really matters to respondents — dropped off rapidly between 1970 and 1972 and decreased steadily, if more slowly, after that. Essentially, the polls show that people cared, but not as much as they cared about other social issues. If forced to choose, people said they would choose "economic growth" and "adequate energy" over environmental protection. Dunlap concluded that support for the environment was "widespread but not terribly intense," and that such a shallow, " 'permissive consensus' . . . [does] not translate directly into pro-environmental votes and political action" (1987:36).

Measured by the yardstick of legislative or electoral results, pro-environmental opinion apparently did lack depth or staying power. True, the 1970 peak in environmentalism had produced a flurry of legislative activity, but corporate interests subsequently found it rather easy to convince the public that "too much regulation" was to blame for economic problems. By 1980, the champions of deregulation had gained the upper hand and were triumphantly undoing the regulatory gains of 1970.[7]

But that is not the whole story. Issue salience is tricky and complex. It may vary in ways not detectable by opinion polls. It may show up in places other than the voting booth.

People had become much more aware of air and water pollution and of the potential adverse health effects of industrial chemicals. Continuing bad environmental news heightened their sense of being at risk. Even if such environmentalist beliefs were not continuously salient in most people's minds and hence did not appear as voting behavior or other formal political activities, they could persist in a latent state, ready to burst into salience if a local threat provided an experiential basis for immediate concern. Problems that previously had been either poorly understood or thought merely a nuisance, problems that had been either ignored or fatalistically accepted, could now more easily be construed as threats that required protest and mobilization.

Research shows, in fact, that at the same time that environmental concern seemed to be declining as a *national* political force, it was very much alive on an entirely different level, vigorously manifesting itself in the form of *local* organizing, local conflict. The Environmental Conflict Project, for example, found that thousands of instances of local environmental conflicts occurred in the United States during the 1970s.[8] Although most of those conflicts occurred in a handful of highly industrialized states, there was a marked geographic diffusion trend that saw conflict spread "from frost belt to sunbelt" (Gladwin, 1987:19). These conflicts tended not to be about nature, per se, but about "land use, social impact, human health" (ibid.:23).

Some of these local conflicts were about industrial wastes. In some cases, people became convinced their communities had been contaminated by toxics. In their excellent review of the literature on contaminated communities, Kroll-Smith and Couch provide a compelling description of the experience of contamination, the stress, the uncertainty, the "chronic state of dread," "the disrup[tion of] the routine relationship between people . . . profound social and psychological trauma" (1991:ii, 2). The attitudes and beliefs promulgated by modern environmentalism increased the likelihood that unpleasant but ambiguous conditions would be construed as evidence of contamination. And, although Kroll-Smith and Couch say that "comparatively few people . . . respond [to contamination of their community] by becoming politically active" (ibid.:24), cases cited in the literature confirm that, in the 1970s, the belief that one's community was contaminated was beginning to provoke, here and there, an activist response.[9]

In other places, people organized when they found out that a hazardous waste facility, a chemical landfill or an incinerator, was planned for their community. There is a long history of community resistance toward the siting of what Popper calls LULUs, locally unwanted land uses—facilities and infrastructural construction projects (halfway houses, prisons, housing projects, hospitals, airports, highways, power plants, and the like) that provide a good for society as a whole but concentrate all the undesirable costs in the host community.[10] Although the 1973 EPA survey cited in chapter 2 suggests that, at that time, people still viewed the prospect of having a toxic waste facility for a neighbor with equanimity, by 1976, 42 percent of TSD owners and operators were reporting that public opposition had become a problem for them, and in 1979 EPA's consultants had no trouble retrospectively finding numerous instances of community opposition to facility siting.[11] Clearly, the beliefs propagated by modern environmentalism had begun to change the public's view of hazardous waste facilities, from an attitude of neutrality toward the view that such facilities are LULUs that communities ought to oppose. (Both types of local hazardous waste action, protests in contaminated communities and siting opposition, will be discussed at greater length in chapter 4.)

Breakthrough to Issue Formation, 1978-80

The proliferation of local actions was an important moment in the transformation of hazardous waste from a nonentity to a full-fledged Issue. Such protests were ideal candidates for attracting news coverage. Sociological studies of television news have described television's preference for "disorder" stories that feature disasters, victims, protesters, and its stylistic preference for stories with plenty of action and color, as well as for stories where abstract social issues can be personalized, thus made more accessible.[12] Contamination protest was made for television. All the right elements were there: industrial chemicals, cancer and birth defects, victimization of innocent citizens. In addition, hazardous waste struggles were rich in possibilities for great visuals: sinister piles of drums, discolored pools of water, angry community meetings, distraught parents.

Each protest was a potential candidate. It would take only one case among the many to break out and become the symbolic kernel of a national opinion-formation process; only one grain of sand around which the perceptual pearl would begin to form. The breakthrough came, as we know, with the case of Love Canal and, to a lesser extent, a handful of similar contamination events. The story of Love Canal has been told often and told well by participants, journalists, environmental activists, members of Congress, and sociologists and other social scientists.[13] I do not wish to reinvent this wheel and, in any case, the specific events at any one locale are not the point. Moloch and Lester, in their paper on media coverage of the Santa Barbara oil spill,[14] made the important distinction between how an event is experienced locally and how that event is represented in the national media. My interest in Love Canal, then, is not the local story, but the form of its appearance to a national audience, "Love Canal as Seen on TV." The question is, What did the nation see and hear? One has to bracket the local news story and look at Love Canal only as seen on television and in the popular press. One has to bracket the local political struggle and examine how Love Canal was appropriated and represented in political discourse in Washington, D.C.

Hazardous Waste on Television

The nation first heard about hazardous waste when CBS and ABC carried stories about events at Love Canal, on August 2, 1978. Over the next two years, the networks' nightly news programs alone carried about 190 minutes of news about hazardous waste.[15] Whether people watched television in the morning (*The CBS Morning News, Today, Good Morning America*), during the day (*The Phil Donahue Show*), the dinner hour (*The MacNeil/Lehrer Report*, the networks' nightly news shows, *60 Minutes*), evening (*Nova*), or late night (*Nightline*), they saw stories about Love Canal.[16] I acquired copies of fifty-one stories that, together, accounted for about two-thirds of the time, 120 minutes, devoted to this topic by the networks' nightly newscasts and carefully examined them in order to extract from them their most characteristic qualities, their most repetitive verbal and visual messages.[17]

Love Canal's First Appearance on Television. During the first month of coverage, reports provided the basic background information. Decades earlier, Hooker Chemical had dumped toxic chemical waste, "thousands of drums" (C1) of it, into Love Canal, a partially completed and abandoned navigation channel. In 1952, the canal was covered up. A year later, Hooker sold the land to the Niagara Falls Board of Education. A school was built. Developers built homes and "unsuspecting families" (C1) moved in. In the 1970s, after heavy rains, chemical wastes began to seep to the surface, both on the school grounds and into people's yards and basements. Federal and state officials confirmed the presence of eighty-eight chemicals, some in concentrations 250 to 5,000 times higher than acceptable safety levels. Eleven of these chemicals were suspected or known carcinogens; others were said to cause liver and kidney ailments (C1, A1, A6).

The sense that this was a disaster was conveyed explicitly: a "calamity that has been steadily and silently building up for a long time" (A1) was now "seeping . . . coming out of the ground" (C1) into people's lives. More subtly, the reports conveyed the ominousness of the events at Love Canal by showing visuals that seemed to signify "normalcy," but undermining or reversing, signifying the opposite, through voice-over narration. Some examples: A boy bicycles along a quiet suburban street while the narrator says, "There have been instances of birth defects and miscarriages among families" (C1). Kids play on a community playground while the narrator reports the New York Health Department's recommendation that pregnant women and small children evacuate (C1). Aerial views of beautiful Niagara Falls and a boatful of admiring sightseers are offered as the narrator quotes officials' fear that wastes are contaminating the Niagara River and Lake Ontario (N1). Shots of people's backyards, lawns, and swimming pools are followed by pictures of holes in the ground filled with ominously colorful soups of liquid chemicals. "More than eighty chemical mixtures had worked their way up through the earth . . . into nearby backyards and basements" (N2), ruining swimming pools, "destroying much of the vegetation" (C1). This device, often found in horror movies, conveys a sense of insidiousness, of benign everyday life suddenly becoming nightmare.

Known and suspected health impacts were by far the most important aspect of the story. High rates of miscarriage and birth defects were reported (C1, C2, A1-A4, A6, N2). "Fear and concern [are] gripping this community" (A1). Lois Heisner, a housewife and former Hooker employee, describes two miscarriages, two children born prematurely, one "has a defect" (A1). Karen Schroeder, interviewed with her daughter Sherry at her side, a little girl with large, sad eyes, says her daughter was born with a cleft palate and two rows of bottom teeth, and that she has congenital hearing problems (A6; also shown and described in C1). She says, "Five children are retarded out of 24 . . . too much to be a coincidence" (A6). "You could see the fear on faces," says the reporter as the camera shows residents lined up to have their blood tested for "abnormalities" (C2). The state health official administering the blood test tells the reporter that they are looking for evidence of abnormalities associated with leukemia, anemia, toxic liver conditions. He says, "I would expect to find more than the normal level of these kinds of illnesses" (C2). Indeed, CBS reports that, "of those who have been told [the results of their blood tests], abnormalities have shown up in 40 percent of the cases" (C5). It follows with an interview of a housewife worried about leukemia and a report of findings of abnormalities in some residents' liver function. Another mother, Jean Guagliano, tells the reporter that her children both have kidney disease. The reporter from ABC sums up the story as an "environmental disaster. . . . people who lived here . . . [are] damaged for life" (A6).

The families are shown grappling with financial tragedy as well. If they are ill, the reporter says, they will have to pay for medical treatment themselves (C5), while the visuals of working-class or lower-middle-class homes convey the message that most are barely making ends meet and have no way to bear major medical expenses. Worse, their one major source of savings, their homes, suddenly have no market value. A reporter asks a young woman, Mary Heeney, "Have you thought about trying to sell [your home]?" She laughs bitterly. "They won't put for sale signs on any of these homes and who'd be crazy enough to buy one? I can't imagine why anybody would" (C2). Health and financial burdens interact. They want to get out, to flee the hazard, but how can they afford to go? Lois Heisner, holding her child, says, "We have decided we are going to get out one way or another, but right now, you know, you can't just jump; where are we going to go?" (C1).

Stories conveyed the total disruption of community, of settled, everyday life. The grade school is closed; the school yard, located on top of the canal, "where children have played for twenty-four years" (N2), is fenced off and signs on the fence warn, "No Trespassing," "Dangerous Area." Officials recommend that people most affected should evacuate the area, first 37 families with pregnant women and children under 2 years old, then 97 families, eventually 239. As families begin to relocate to temporary housing, there are shots of boarded-up homes. One boarded-up house still sports a protest sign, "Love Canal Recipe: 1 Mix 82 chemicals 2 place in canal for 25 yrs. yield: sickness and death" (N2). But, at least, these folks *are* getting away; families who have not yet succeeded in getting officials to relocate them are angry. They protest that it is "dangerous for them to remain along Love Canal" (C2). Their houses also have protest signs: "We want out now" and "Evacuate us *now!*" (C2). The cameras show children demonstrating along a roadway, holding signs: "Children of *all* ages should be evacuated"; "Love Canal stinks and kills" (C2).

The most frequent, most persistent images throughout these news stories were of community lands (school yard, suburban field, backyards) that *ought* to be green, vibrant with suburban/domesticated vegetation, but instead show only sparse, half-dead plant cover, punctuated with holes filled with unnatural-looking chemical soup; house yards and basements invaded by chemical ooze; disrupted neighborhood life, a closed school, boarded-up homes, mass blood testing; families packing and moving; protest signs on houses; worried housewives and mothers. The stories contained nine interviews with residents, all women, all working-class or lower-middle-class, all articulate, competent, smart, some angry, all worried about the health of their families.

These early stories also began to suggest that what happened to the people at Love Canal could well happen in many other communities, that the horror movie of chemicals coming out of the ground into your home and your children's bodies could soon be playing at a theater near you. EPA officials said that waste "has been disposed in unsafe ways" and that dumps sites are "like ticking time bombs" (C4). The EPA was said to have found at least 400 other cases (C4) or 638 cases (A6), that there may, in fact, be thousands (A6). U.S. industry was said to be generating thirty to forty million tons of waste that "could have a severe adverse effect on human and animal life" (C4). Several stories emphasized that safe disposal methods existed, but strong federal regulations (i.e., RCRA) would not go into effect until 1979 (C4, N2). One reporter noted that, for now, "proper disposal of chemicals is the exception, not the rule," and repeated EPA's estimate that 80 percent of waste was being disposed "improperly, inadequately, or illegally" (C4).

Other Instances, Other Victims. Walter Cronkite, one of the most trusted men in America, introduced one story with these words: "The problem of hazardous chemical waste drifts across this nation these days like a bad dream" (C6). Indeed, in the two years following initial coverage of Love Canal, television news featured a steady stream of stories that, taken together, communicated that toxic chemical contamination was widespread:

> PCB wastes are illegally dumped along roadways in North Carolina (C4, N1). PCBs also are illegally dumped in Ellisville, Missouri (A9). PCBs are found in a creek that runs into the Niagara River (A5).

> Pesticide residues from a Velsicol dump contaminate domestic water supplies in Toone, Tennessee (Hardemann County), forcing residents to rely on trucked-in water. A resident suffering from strokelike symptoms says, "I think it's killing us" (C6).

> At a bankrupt facility in Lowell, Massachusetts, 15,000 drums of waste are found (C8).

> Thousands of drums of "poisonous chemicals" are found dumped at Stump Gap Creek, Kentucky, the famous "Valley of the Drums" (C10).

> Five pounds of dioxin cause an environmental crisis at Seveso, Italy; a ton of dioxin is dumped by Hooker Chemical at the Hyde Park dump in Niagara. Dioxin is found in Bloody Run Creek. Members of a family living near the creek all have respiratory problems. Fifteen workers at a factory next to the creek have had cancer (C11).

> Mercury wastes are dumped at a site in Woodbridge, New Jersey, the "worst case of mercury contamination in the world today." Citizens are having blood and hair tested to assess level of contamination (C13).

> In one story, Hooker Chemical is said to have contaminated groundwater and White Lake, near Montague, Michigan, with C-56, a deadly compound, and eight other cancer-causing chemicals; knowingly polluted the air and discharged fluorine and phosphorous into Spring Creek at White Springs, Florida; dumped asbestos and chlorine at Taft, Louisiana; violated discharge permits at Columbia, Tennessee; and contaminated groundwater with cancer- and sterility-causing pesticide residues at Lathrop, California (C18).

> Neighbors of the Hollywood Boulevard dump in Memphis, Tennessee, complain of illnesses and contaminated farm animals. On one block, "every household claims some chemically related ailment, from cancer, to skin rashes, to birth defects." Officials say that they have not been able to locate a source that could account for the problems; one woman who has had cancer charges "coverup right from the top . . . right on to the bottom" (C23).

In Woburn, Massachusetts, there is a cluster of cancer cases. The news shows a boy struggling with leukemia; he has lost his hair from chemotherapy treatments (C25).

Livestock in Michigan are contaminated by PBBs, possibly leaking from a nearby disposal site (A9).

A chemical waste storage warehouse in Elizabeth, New Jersey, explodes and burns on Earth Day, 1980 (N3).

Some 50,000 tons of acidic sludge threaten a wealthy community in Fullerton, California. People have difficulty breathing, suffer from nausea. A woman resident says, "We should try to, to avoid what happened at New York; another Love Canal" (A9).

Love Canal Back in the News. Television checked in at Love Canal periodically. Residents who left reported that their health improved (C11). Those who remained felt trapped, desperate to leave but unable to sell their homes. They were shown protesting, chanting, "We want out" (C12).

Then, in May 1980, Love Canal once again became *the* focus of attention, with eight stories over five days. The EPA released a study that found that eleven of thirty-six residents tested had chromosomal damage (C26, N4). A doctor from the EPA, the scientist who had conducted the study, and a cancer researcher working with the community all stated for the camera that the kind of chromosome damage found could cause cancer, miscarriage, and birth defects (C26, N4, A7). Although the stories noted uncertainties and questions about the findings (A7, C27, C28, N6) and showed Hooker Company representatives denouncing the findings as "preliminary," "inconclusive," and "inadequate" (C26, N4), interviews with distraught citizens were clearly the emotional core of the coverage. In response to a reporter who asked, "What do you think about when you think about tomorrow, next month, next year?" Pat Sandonato, one of the eleven found to have chromosome damage, responded, "My kids dying, coming down with something they can't cure" (C26). Phyllis Whitenight, another of the eleven, said she has had cancer and "will probably get it again" (N4). Still another, Barbara Quimby, said she was not surprised to hear the findings, as her 8-year-old had birth defects and was retarded (N4). Marie Pozniak, who has had cancer and whose daughter has central nervous system problems and asthma, said, "Every child I look at in this area, I wonder, 'Do you have hidden dangers in you?' How many cancers are growing in their body?" (N6).

The residents who were not part of the original relocation had been fighting for almost two years to get the government to buy their homes and relocate them, too. The chromosome study raised their sense of urgency and rage to new heights. The reporters found them "furious" and "desperate to move" (N4). They demanded that the government "evacuate them immediately" (N5). Leonard Whitenight, the husband of the woman who feared she would get cancer again, raged that the government has money for Cuban refugees or to get bodies out of Guyana, but "won't help its own people" (C26).

The cameras showed that everyday life had come to a halt. People were in an uproar, standing in nervous groups, milling about in the streets. Barbara Quimby said, "The people are just at their wits' end. They just can't handle any more mentally. Not when it comes down to your children. That's the worst" (A7). The following day, citizens held two EPA officials hostage for five hours. Barbara Quimby, holding a small child, said that "the people even scared me. They're my friends, my neighbors. They're calm, law abiding citizens, but they just can't handle it anymore. It's not only the risk to you. When you hear that this could affect your children. I mean, anybody who's a parent has to sympathize with us" (C27).

The following day, Love Canal was on all three networks, the lead story on two of them (C28, A8, N6). President Carter declared an emergency. Temporary relocation of 700 additional families was announced. The newscast showed mixed reactions, some celebration and relief, but also dissatisfaction. People protested that temporary relocation to motels would not solve their problems. Roy Cleveland, a man whose wife had cancer and whose children had birth defects, said it was "better than nothing, but not a hell of a lot better" (C28). Lee Lutz said, "We just want permanent get out of there [sic]" (A8). Some declared they would stay until the government provided a permanent solution, but most went and went quickly. One mother, her family already in the car, told a reporter that her nephew has cancer and she "can't put my children through that; I can't take the chance" (N6). Pat Sandonato, the one who said she thought about her kids dying, said, "I just want my kids out of here" (C28). Another family was shown leaving just two and a half hours after the announcement. The kids said good-bye to their pet, a Saint Bernard, left behind as they drove off to a motel (A8).

Scope Stories. Coverage of catastrophes in specific communities was accompanied by numerous stories about the overall scope of the problem nationally. The figures painted an ominous picture. The *known* scope of the problem was huge already and, clearly, the full extent of the problem had yet to be discovered. The EPA was reported at various points to have estimated that 638 sites needed attention (C7); that "more than 800" represented a serious threat to public health (C19); that there may be "as many as 2,000" that would need Superfund cleanups (C14); that there may be as many as 30,000 sites with some waste, but not enough was known about them (C7, C19, A10). In 1978, it was reported that industry was generating 35 million tons of hazardous waste a year (C8); in 1980, the estimate had been doubled, to "125 billion pounds" (A10). EPA's estimate that 80-90 percent of this was being disposed in unsafe ways was reported over and over (C4, C17).

Industry; Government Officials. These amounts might have been less frightening if the news had shown corporate generators and the disposal industry as acting in a responsible manner and if government regulators had been shown to be competent to protect public health. Viewers, however, got just the opposite impression.

EPA officials repeatedly condemned past industry practices, saying that industry had left the nation a legacy of vast levels of contamination. In a story that identified Hooker Chemical as a chronic environmental scofflaw, a Hooker official admitted that "with regard to the things that have gone on in the past, where we're judging what we did twenty or thirty years ago by the standards of today, ahh, we, ahh, we have problems" (C18). He insisted that things had improved, but that is not the impression conveyed by the news. The chemical industry was shown resisting stronger regulation. Ralph Nader charged that members of Congress who had been delaying passage of Superfund were getting big campaign contributions from the chemical industry (C31). Cameras showed a roomful of industry lobbyists at a Senate Finance Committee hearing on Superfund as Congressman Gore charged that "the chemical industry is engaged in a very cynical effort to try to kill this legislation" (A9). When the Carter administration finally proposed implementing standards for RCRA, a Union Carbide spokesman complained, "There's too many regulations [sic], with standards that simply spend money" (C35) and a DuPont vice president called the proposed standards "silly" (A10). As RCRA was about to go into effect, Walter Cronkite reported that anticipation of new, stricter, more costly disposal rules had "triggered a vast, illegal dumping." There were pictures of tanker trucks clogging the road as the reporter said, "As the midnight deadline approached, there were reports of massive amounts of toxic chemicals on the move around the country, all in an effort, according to EPA officials, to dump the wastes before the new reporting requirements take effect" (C35). This was hardly the kind of news that would reassure worried citizens.

Could government provide the needed protection? Not likely, according to the network news. Individually, local, state, and federal officials are generally shown in a positive light. They care; they carry out health surveys and do environmental monitoring; they announce various forms of assistance for affected citizens and assist in temporary relocations. But regulatory bureaucracies, as a whole, fare less well. If safe disposal methods exist (C4, N1) and strong regulations have been on the books since 1976, why is 80 to 90 percent of waste still improperly disposed? Implementation has been delayed, EPA's enforcement efforts are weak, "infinitesimal in comparison to the size of the problem . . . grossly inadequate," according to a spokesman for the Environmental Defense Fund (C19). When the EPA finally promulgates RCRA standards, stories note that standards for disposal facilities are incomplete and that many hazardous substances have been left off the list and will fall outside the system (C35). EPA is said to have only eighty inspectors to monitor compliance at 20,000 disposal sites and 30,000 factories (C35). "Without a larger enforcement staff," says the reporter as he concludes a story on RCRA finally going into effect, "few expect that the new law will quickly clean up the toxic waste problem" (C35).

TV Iconography of Hazardous Waste in a Nutshell: A Forty-Second Précis. By November 1980, television's visual and rhetorical vocabulary of hazardous waste had become so highly conventionalized that Rebecca Chase could show it all and tell it all in forty seconds of snapshots and sound bites (A10):

Visuals	*Voice-over*
Boarded-up homes	[No narration. Clicking of a slide projector carousel advancing to the next slide.]
Chemical plant	
Polluted, dead ground	
Boarded-up homes	"Love Canal was just the beginning.
Chemicals leaking from container	"Deadly chemicals seeping into basements
Backyards of boarded-up homes	"and backyards from an abandoned dump.
Sign: "Hazardous Chemicals Unauthorized Entry Prohibited"	"A national disaster declared.
Boarded-up homes	"Residents evacuated.
Drums of waste	"Once officials started looking,
Truck discharging liquid waste	"they found toxic chemicals throughout the nation,
Drums	"improperly stored . . .
Drums	"abandoned . . .
Polluted water	"contaminating drinking water . . .
Caterpillar tractor putting dead cattle in a mass grave	"killing cattle . . .
Person in a wheelchair	"making people sick . . .
Child from Woburn, MA	"cancer . . .
Woman	"miscarriages . . .
Child from Love Canal	"birth defects.
Aerial view of a dump site	"The EPA has discovered 30,000 dumps . . .
Workers sampling from drums	"4000 are considered hazardous.
Houses	"From New York . . .
Drums	"to Kentucky . . .
Drums	"to Illinois . . .
Polluted water	"to Louisiana . . .
Drums	"to California; toxic time bombs.
Explosions at Elizabeth, New Jersey	"Some have already blown up."

Hazardous Waste in the Popular Press

Although television is the critical medium today for attitude formation, the process becomes only more powerful if the message is amplified by being conveyed in other media as well. Love Canal stories appeared in almost every type of popular magazine, the newsweeklies (*Time, Newsweek, U.S. News & World Report*), the business press (*Business Week, Fortune*), the left/movement press (*The Nation, The Progressive*), science journals (*Science*), highbrow literary magazines (*The Atlan-*

tic, Saturday Review), women's magazines (*Redbook, McCall's, Glamour*), and other outlets such as *Mechanix Illustrated, People*, and *Reader's Digest.*

The sheer level of news saturation is, itself, an important datum. By 1980, one would have had to have been militantly uninterested in the world, uninterested in reading *anything* (on the order, perhaps, of people who could serve on Oliver North's jury), in order to avoid reading something about Love Canal and hazardous waste.

I examined the stories that appeared in *Time, Newsweek, Redbook, McCall's, Glamour*, and *Mechanix Illustrated.*[18] How was hazardous waste represented in the popular press? What impression did people get when they looked at these stories?

It all depended on *how*, how deeply or how carefully, the reader encountered these articles. A careful reading of the articles conveyed a frightful story, certainly, but the articles also tried for balance and fairness and the reader was presented opposing views. If, however, these articles were glanced at quickly and superficially, their message was one-sided, highly stereotyped, top-heavy with charged, inflammatory rhetoric and imagery.

The careful reader could have read beyond the overheated rhetoric—"witches' brew," "nightmare," "environmental disaster"—and still been shaken up by the details, the purported health problems, the disruption of those poor people's lives. Officials' statements—"We don't know where they are. We don't know in what quantities" (*Newsweek*, August 21, 1978:26) and "even an extraordinary effort, commenced immediately, cannot achieve adequate protection for the American public for years to come" (*Newsweek*, October 22, 1979:51)—conveyed the message that the nation was just waking up to a disaster of unknown dimensions, that this could and would happen to others. The careful reader would have also learned, however, that there were substantial doubts. It was reported that the chromosome damage study at Love Canal may have been flawed. More generally, scientists had yet to prove conclusively that people's health problems had indeed been caused by proximity to waste dumps.

One would, however, have to have read these stories with extraordinary detachment to give these reservations equal weight. We know, furthermore, that many "readers" glance rather than read. They look at the pictures, read the captions and headlines, perhaps the first paragraph, and then turn the page. Such glancing, superficial readings produce an extreme, very one-sided message.

The photographs that appeared with these stories faithfully repeated the limited, highly stereotyped, emotionally charged visual vocabulary of television's "toxic waste" imagery: haphazard piles of broken, leaking fifty-five-gallon drums; cleanup crews encased in protective safety gear; home after home, boarded up, abandoned; plain folks, mostly women, distraught, angry.[19] Following are some captions that accompanied these pictures:

Environmental disaster area: a witches' brew of long-buried chemicals turns the Love Canal district into ghost town. (*Newsweek*, August 21, 1978)

Ending a nightmare: Jo Ann Kott's family packs up and heads for a motel. (*Newsweek*, June 2, 1980)

An unexpected eruption at the Three Mile Island of waste dumps. (*Time*, May 5, 1980)

"You don't want them near you—nor do I." (*Time*, September 22, 1980)

"We didn't understand that every barrel stuck into the ground was a ticking time bomb." (*Time*, September 22, 1980)

Incessant medical tests are part of the tragedy of the Love Canal disaster. (*Mechanix Illustrated*)

The Quimbys—Barbara, Brandy, James and Courtney—even their dog—have probably been breathing poisoned air all their lives. (*McCall's*)

A high incidence of cancer, birth defects and respiratory and neurological problems. (*Time*, September 22, 1980)

Today it is a ghost town of empty houses and abandoned dreams, but the people of Love Canal have an important story to tell, for every day it becomes clearer that what happened to them could happen again—to any of us. (*Redbook*)

A vivid illustration of the problems facing many communities throughout the U.S.: thousands of drums containing toxic wastes. (*Time*, September 22, 1980)

Finally, these are some of the articles' headlines, printed big, in boldface:

A Nightmare in Niagara (*Time*, August 14, 1978)

The Poisoning of America: Those Toxic Chemical Wastes (*Time*, cover story, September 22, 1980)

The Tragedy of Love Canal (*Redbook*)

Underground Time Bombs: Chemical Waste Dumps (*Mechanix Illustrated*)

Our Fear Never Ends (*McCall's*)

Media Coverage Stimulates Politicians' Rhetoric

Before the events at Love Canal, both the Carter administration and Congress treated hazardous waste as unimportant, a nonissue. Congressional oversight hearings in 1977 and 1978 focused almost exclusively on the solid waste aspects of the law. The attention of Congress was on oil spills and spills of other hazardous substances, rather than on contamination from what would later be labeled the "inactive hazardous waste site problem."[20] In the absence of congressional scrutiny or serious outside political pressures, the administration was neither eager nor swift to implement RCRA's hazardous waste provisions. RCRA implementation was subordinated to other administration goals. Draft regulations were weakened and delayed, fledgling enforcement actions were stifled internally, the program was underfunded.

Media coverage's wonderful power to attract and focus politicians' attention was again demonstrated by events following the transformation of Love Canal from local to national event in August 1978. After that, both the administration and Congress hurried to get in front of the issue, to appear to be leading the effort to make new policy to deal with this most alarming development.

Congressional oversight activity intensified. The focus of the oversight hearings shifted from solid waste, originally seen as the most important part of the act, to the hitherto neglected hazardous waste provisions. Congress scored the EPA for failing to meet statutory deadlines, failing to get implementation off the ground, weak enforcement, and other failings.[21] Members of Congress competed to lead the call for new legislation to speed the cleanup of abandoned waste dumps.

Coverage of Love Canal also signaled fundamental changes in the Carter administration's approach to hazardous waste. Locally, at Niagara Falls, the White House was caught between conflicting imperatives to avoid political embarrassment and to avoid incurring major, and precedent-setting, financial obligations; it hesitated, providing relief to Love Canal residents reluctantly and only when the potential for embarrassment outweighed other considerations.[22] Although the administration was rather cautious and reluctant when dealing with Love Canal as a local and specific case, nationally, its response was markedly different. It seized the initiative on the "orphan" site problem, declared Love Canal only the tip of the iceberg of a vast environmental crisis, and made the "Tragedy of Love Canal" the rallying cry for passage of what would come to be known as the Superfund.

The administration's and Congress's efforts to get out in front of a fast-breaking story contributed to transforming Love Canal into *the Tragedy of Love Canal* and hazardous waste into *the nation's Most Important Environmental Problem*. Love Canal was repeatedly, ritualistically, termed a public health "tragedy."[23] Some claims about tragic health impacts were cautious. For instance, a House committee report noted, "The Love Canal health data shows elevated miscarriage and birth defect rates; evidence suggests many other health effects the nature and extent of which are in dispute" (Senate, 1983, vol. 2:50). Most statements were, however, much more categorical, much less likely to note continuing uncertainties. They boldly asserted the certain presence of a much larger number of serious conditions:

The residents of Love Canal suffer from a variety of severe, crippling diseases. (Senator Stafford, in Senate, 1983, vol. 1:59)

Severe health problems discovered at Love Canal. (Senator Randolph, in ibid.:684)

High incidences [*sic*] of birth defects, miscarriages, skin, liver, and kidney ailments, epilepsy, and depression. (Senator Culver, in ibid.:147)

Birth defects, miscarriages, epilepsy, liver abnormalities, sores, rectal bleeding, headaches—not to mention undiscovered but possibly latent illnesses. (Michael H. Brown, in *New York Times Sunday Magazine*, cited in ibid.:316)

Next, politicians repeatedly stated that "Love Canal is not unique," that there are many Love Canals. Two rhetorical devices were used to buttress this claim. First, the names of a handful of other contamination episodes were repeated constantly, almost ritualistically. About a dozen other cases of local contamination from improper disposal were cited by name during congressional action on Superfund. Of these, three were cited with some regularity: the 17,000 barrels of waste found dumped at the "Valley of the Drums," near Louisville, Kentucky; contamination of water supplies by the Velsicol dump in Hardemann County, Tennessee; and contamination of the Cedar River near Charles City, Iowa.[24]

Second, the impact of naming other known contamination cases was amplified by citing increasingly ominous estimates of the full scope of the problem. A joint EPA and Justice Department task force reported to the Senate that it had "identified 5,790 hazardous waste sites which were to be investigated" (Senate, 1983, vol. 1:311). An EPA survey of

> pits, ponds, and lagoons used to treat, store, and dispose of liquid wastes identif[ied] 11,000 industrial sites with 25,000 such surface impoundments. At least one-half of the sites are believed to contain hazardous wastes. The survey found virtually no monitoring of ground water. . . . 2,455 of the 8,221 sites assessed are unlined, overlie usable ground water aquifers, and have intervening soils which would freely allow liquid wastes to escape into ground water. (ibid.:692-693)

The Hart study, also commissioned by EPA, estimated that the nation had 30,000 to 50,000 sites containing some hazardous wastes, with perhaps 1,200 to 2,000 of those presenting "potentially significant problems" (ibid.:61). Members of Congress and EPA spokespersons concluded:

> Love Canal is not a unique problem. (Senator Randolph, in ibid.:684)

> Unfortunately we learned that Love Canal was not a unique event. (Congressman Florio, in Senate, 1983, vol. 2:226)

> Tip of an iceberg. (Congressman Long, in ibid.:216)

> After Love Canal, an explosion of similar incidents burst forth across the nation. . . . We learned that Love Canal was merely the first detonation of a string of chemical time bombs literally strewn across the nation. (EPA head Douglas Costle, in Costle and Beck, 1980:425)

If Love Canal was a Tragedy and there were many, many potential Love Canals, chemical time bombs silently ticking, soon to detonate, improper disposal of hazardous wastes was surely one of the, if not *the*, most important environmental problems in the United States:

> Some of the most significant environmental and public health problems fac-

ing our Nation. (President Carter, in message accompanying proposed legislation, in Senate, 1983, vol. 3:25)

One of the most pressing environmental problems confronting our society.
(Senator Culver, introducing S.1480, in Senate, 1983, vol. 1:147)

Has become one of our Nation's most serious environmental problems. (Senator Kennedy, during Senate floor debate, in ibid.:765)

Perhaps the most serious environmental problem facing the Nation today.
(EPA administrator Jorling, testifying for the administration, in ibid.:60)

What many call the most serious health and environmental challenge of the
decade. (Senate committee report to accompany S. 1480, in ibid.:309)

Fast becoming the most serious environmental problem of our time. (Senator
Chafee, during Senate floor debate, in ibid.:765)

Mr. Chairman, improper hazardous waste management is the most serious
environmental problem facing our Nation today. (Congressman Florio, opening floor debate on H.R. 7020 in the House, in Senate, 1983, vol. 2:225)

Even the name, Superfund, was part of this rhetorical storm: BIG legislation, a Superfund for a Super Problem.

Media coverage and political rhetoric fed on each other. Coverage was what had
stimulated intensified political activity in the first place; politicians' rhetorical
moves subsequently became part of, helped frame and define the meanings of, the
news stories. EPA administrators were on camera, graphically explaining the problem, predicting that many more cases of contamination would be found (C4, N1,
A10). The news carried the agency's ever-growing, ever more ominous estimates of
the number of sites that were out there, ticking, waiting to explode into people's
lives.[25] Television conveyed President Carter's feeling that his Superfund proposal
was his "most important" environmental initiative (C20). As the Superfund bill
worked its way through Congress, television featured key congressional leaders,
Senators Kennedy and Moynihan, Congressmen Moffett and Gore, all repeating the
same basic message: Love Canal a Tragedy, many Love Canals, most important environmental problem.

Polling Confirms That a New Issue Is Formed

Opinion polls soon confirmed that all this—the media attention, the politicians'
rhetoric—had had a big impact on people's perceptions. An ABC News/Harris poll
fielded in June 1980, shortly after coverage of the Love Canal chromosome damage
study (May 17), the taking hostage of two EPA officials (May 19), and the president's declaration of a federal emergency at Love Canal (May 21), found that 93
percent of the public favored making federal disposal standards "much more
strict"; 86 percent favored making "toxic chemical dumps and spills a very high
priority for federal action"; 88 percent favored "providing funds to allow people
who live in such areas and whose health may have been impaired to move out as

soon as possible."[26] A poll commissioned by the Chemical Manufacturers Association found that 93 percent of politically active respondents felt either "very" or "extremely" concerned about chemical industry waste disposal practices.[27] A third poll, conducted by Robert Cameron Mitchell for the President's Council on Environmental Quality, showed that 64 percent of the public was concerned "a great deal" about chemical waste disposal.[28] It was this poll that yielded the "how close is close enough" findings reproduced in Figure 1 in chapter 2.

Glamour magazine devoted its regular reader poll, "Tell Us What You Think About . . . ," to Love Canal in August 1980.[29] The very fact that such a poll found its way into a glossy fashion magazine suggests the breadth and diffusion of issue awareness. Although this poll was methodologically flawed—no sample size is reported; it is impossible to evaluate the effects of respondent self-selection—and therefore of questionable reliability, it told the same story. A total of 92 percent of the respondents said that news reports about Love Canal had made them "more alert to the dangers of living near a waste site"; 86 percent said the government was not doing enough to protect the public; 98 percent favored new legislation "to prevent dumping problems in the future."

The polls not only confirmed for all participants that the issue-creation process had worked; their findings were fed back in to process, became part of the process, further augmenting and legitimating it. Poll results were themselves news, and the public was duly informed about its collective opinions. Opinion survey results were also fed back into the legislative process. Senators and members of Congress referred to them during floor debate to remind their more recalcitrant colleagues that Superfund legislation enjoyed overwhelmingly popular support.[30]

Once Formed, the Issue Persists

It should be noted, finally, that, having formed, the images and dreaded meanings associated with "toxic waste" have proven quite durable. Hazardous waste is still seen as an extraordinary threat. Recently, the EPA conducted a poll to compare how lay people and experts rank environmental problems in terms of seriousness. The experts at the EPA ranked "active" and "abandoned hazardous waste sites" as far less important than other environmental threats; in contrast, the lay public continued to view these as the most serious environmental risks facing the nation.[31]

"Toxic waste" is still associated in the popular mind with images of the ruptured fifty-five-gallon drum and the cleanup worker in the moonsuit, with cancer and birth defects. By the end of the 1980s, the image and meaning of "toxic waste" had become such a widely recognized figure in popular culture that it had secured a place in Hollywood movies. Characters would fall into the Fateful Vat of Toxic Waste and be transformed into Monster and/or into a being with extraordinary powers. This figure appeared not only in midnight cult movies such as *The Toxic Avenger*, but found its way into the biggest, mainstream blockbusters. In *Modern Problems* (1981), a Chevy Chase comedy vehicle, Chase was accidentally sprayed with toxic material, turned green, and acquired psychokinetic powers. By the end of the movie, the changes induced by the toxic exposure had been equated with demonic possession. In *Batman* (1989), a movie that broke all box-office records, Jack Nicholson fell into the Fateful Vat and was transformed from a rather nasty, but human, crime lieutenant into the hideous and maniacally evil Joker. In *Teenage Mutant Ninja Turtles*, the kiddy movie hit of 1990, four ordinary turtles were dropped into a puddle of toxic waste and transformed into mutated, human-sized, intelligent creatures who possessed the martial arts skills of the ninja and the verbal habits and culinary tastes of Valley kids.

Theorizing Perceptual Change in Contemporary Society

Leading Theories of Risk Perception and Issue Creation

In the terminology currently popular in both academic and policy circles, mass dread of some industrial activity is referred to as the problem of lay "risk perception." Perhaps the best known approach to risk perception comes out of a rather specialized subfield of psychological research known as "decision theory." Decision theory has to do with how people deal with uncertain situations, how they decide trade-offs when they have neither adequate information nor a good intuitive feel for the principles of probability. Decision theorists have shown that people use a small number of simplifying "heuristics" when making judgments under uncertainty. When they turned their attention to the question of lay perception of technological risks, decision theorists argued that two of these heuristics, availability and overconfidence, are especially important. *Availability* means that people judge events to be "likely or frequent if instances of it are easy to imagine or recall"; *overconfidence* leads people to have unwarranted confidence in their judgments.[32]

Decision theorists also attempted to study lay risk perception empirically. Ordinary citizens were asked to estimate the risk of dying from each of a long list of different technologies, substances, or hazardous activities. They were also asked to rate the hazards' characteristics: Was the risk voluntary? Was the risk familiar or new and unknown? Were the consequences likely to be severe? Were the impacts short-lived or, instead, chronic or catastrophic? Nonexperts judged something to be risky to the degree that it was "unknown to those exposed and unknown to science, and to a lesser extent by newness, involuntariness, and delay of effect . . . [and by the] severity of consequences (certainty of being fatal), dread, and catastrophic potential" (Slovic et al., 1985:100).

This approach to risk suffers from the theoretical and methodological limitations of its origins in psychological decision theory. The methodology used, questionnaires administered to artificially assembled samples of populations in experimental settings in no way connected to actual social events, is itself a denial of *process*. The findings are necessarily static and ahistorical. The findings are explained in terms of fixed characteristics of human perception, heuristics, or in terms of supposedly inherent and unchanging characteristics of certain technologies—their disaster potential, for example, or their uncontrollability.

The approach tends to underemphasize perhaps the central fact of lay perception, the fact that jumps out at the most casual observer, namely, that mass perception of risk *varies historically*. At some earlier time, the process or technology in question had not been organized as a "risk" in people's perceptions but was tacitly accepted by the public; later, it came to be perceived as threatening, possibly catastrophic. This process of transformation has been documented most extensively for the important case of nuclear power;[33] change in mass perception of industrial hazardous waste followed a strikingly similar trajectory. The decision theory approach certainly tells us *something*. It describes the end result of this process. But, with the exception of its notion of availability, it has little to say about the dynamics of the process itself.[34]

Long before the recent development of interest in studying "risk," sociologists and political scientists sought to understand and describe the issue-creation process. In sociology, advocates of a "social constructionist" approach problematized the purported objectivity of "social problems," thereby focusing attention on the process of "claimsmaking," the process through which society comes to believe consensually that a problem exists. In political science, Cobb and Elder offered an essentially similar description of the process of issue creation in American politics.[35]

These approaches are attractive because they are explicitly concerned with describing process. They are, nonetheless, problematic because they take the form of ahistorical generalizations about social and political process. Ahistorical, timeless generalizations about "social problem construction" cannot capture the historical originality, the qualitative newness of the issue-creation process in the contemporary United States. To be more specific, these approaches continue to emphasize the word-centered production of meaning—with central terms such as *claims, rhetoric*, and *discourse*—at a time when, many now argue, political communication and the production of meaning is increasingly accomplished through images, not words, through visual rather than verbal representation. Words, claims, rhetoric certainly played a role in making "hazardous waste" an issue, but exclusive reliance on word-oriented notions of issue creation would miss the distinctive qualities of an issue that was largely made on and by television.

We need to find a way of thinking about opinion formation that recognizes the distinctiveness of a process that relies more on the image than the word, a process that is more figural than discursive, a process that creates "meanings" in which the cognitive content is underarticulated and is dominated by highly charged visual components. Some in the constructivist tradition have begun to examine this shift from word to image,[36] but this development has been explored most thoroughly in the literature on "postmodernity."

Postmodernity

The term *postmodern* implies periodization and difference. It suggests that society has recently entered a qualitatively distinct and original phase.[37] I do not embrace this view fully or uncritically. The claim that society has entered an absolutely new moment seems too extreme, overstated. Some of the developments that are cited as quintessentially "postmodern" have existed for many decades; conversely, these developments have not completely displaced older, more traditional or modernist ones.[38] Thus, postmodernity seems, simultaneously, to have been with us for a long time and to have not yet fully arrived. Furthermore, as with other academic developments that catch fire and become fashionable, the term has now been used so promiscuously that it has become increasingly hard to pin down what it does and does not signify. David Harvey comments, quite correctly, that the literature on postmodernity is "a mine-field of conflicting notions" (1989:iix). Still, in spite of these problems, this literature has trenchant, illuminating things to say about contemporary society. I believe that its description of changes in the production and consumption/reception of political messages is relevant to rethinking the process of issue creation today.

There is broad agreement that *postmodernity* means, above all else, profound change in the realm of culture. This change is often described, first, as a new fusion of high and low, elite and mass culture:

> One fundamental feature of all postmodernisms [is] the effacement in them of the older (essentially high-modernist) frontier between high culture and so-called mass or commercial culture, and the emergence of new kinds of texts infused with the forms, categories, and contents of [the] culture industry . . . materials they no longer simply "quote," . . . but incorporate into their very substance. (Jameson, 1991:2-3)

Postmodern culture not only derives much of its raw material from the mass-produced output of the culture industry and from advertising, its very existence is unthinkable without the development of a vast infrastructure of mass communications, television and other, increasingly sophisticated, household technologies of media and entertainment. Not surprisingly, theorists of postmodernity say that that new amalgam of elite and mass culture is inseparably bound up with the process of commodification. Postmodern cultural production "impudently [embraces] the language of commerce and the commodity" (Terry Eagleton, quoted in Harvey, 1989:7). According to Jameson, "Aesthetic production today has become integrated into commodity production generally" (Jameson, 1991:4). "In postmodern culture, 'culture' has become a product in its own right. . . . Postmodernism is the consumption of sheer commodification as a process" (ibid.:x).

I hasten to note that "culture" means much more than the self-conscious practice of the visual arts or of literature. Culture must be conceived of much more broadly, as that ensemble of languages, codes, or systems of representation that constitute a "particular way of experiencing, interpreting, and being in the world" (Harvey, 1989:53). Lash helpfully terms this expanded notion of culture as society's "regime of signification" (1990:5). Culture, in this sense, is woven into the very fabric of social life; it is coextensive with and constitutive of social life.

The literature on postmodernity argues that the changes it describes are not restricted to self-conscious cultural production; all of contemporary culture, in the larger sense, is "postmodern" and can be described in similar terms. Postmodernism, Lash says, is a "very idiosyncratic regime of signification" (ibid.). More dependent on images (or, rather, a pastiche or collage of images) than on words,[39] it embraces "ephemerality, fragmentation, discontinuity, and the chaotic" (Harvey, 1989:44). It produces significations "neither univocal nor stable," but "fleeting" (ibid.:51, 59), significations that are flat or depthless, superficial, while at the same time playful, stimulating, even spectacular and euphoric.

Postmodernity, furthermore, means not only change *within* culture, within contemporary society's regime of signification, it also means a fundamental change in the relationship *between* culture and society. Jameson notes:

> Culture is today no longer endowed with the relative autonomy it once enjoyed. . . . the dissolution of an autonomous sphere of culture is . . . to be imagined in terms of an explosion: a prodigious expansion of culture throughout the social realm, to the point at which everything in our social life—from economic value and state power to practices and to the very structure of the psyche itself—can be said to have become "cultural" in some original and yet untheorized sense. (1991:48)

He also refers to this change as "an immense and historically original acculturation of the Real, a quantum leap in what Benjamin . . . called the 'aestheticization' of reality" (ibid.:x).[40] Similarly, Paul Smith says there has been a "hyperextension of interpellative discourses and representations" (1988:138), and Lash talks of a society "whose very empirical reality is largely made up of images or representations" (1990:14).

If culture has grown, dilated, hyperextended, and colonized or fused with society, postmodernity in culture must necessarily have transformed economics, politics, the texture of everyday life, the quality of subjectivity. Take the case of the economy. On the one hand, economic logic, commodification, comes to dominate cultural production. David Harvey says that postmodernity means an "extension of the power of the market over the whole range of cultural production" (1989:62). On the other hand, the economy no longer produces just commodities, but produces, displays, and constructs desire for "commodity-texts" that are some combination of commodity, representation, image, and discourse. According to Paul Smith:

> The classical narrative of production will not suffice as a way of analyzing late twentieth-century capitalism. . . . the term *production* is inadequate or questionable at this point . . . because of the radical changes in capital development in the last decades. . . . production . . . is newly and complexly articulated into contemporary modes of representation and ideological formations. (1988:135-137)

Jameson comments that the "terms, the *cultural* and the *economic*, . . . collapse back into one another and say the same thing, in an eclipse of the distinction between base and superstructure" (1991:xxi).

The literature depicts, also, a fundamental alteration of the texture of everyday life. When the economy no longer produces only commodities, but commodity-texts, people no longer consume just things, but thing-images. When "everyday life [is] increasingly comprise[d of] representations" (Lash, 1990:12) consumption of representation and image must increasingly become the substance of everyday activity. The distinction between real and representation becomes meaningless. The "world [is] transformed into sheer images of itself . . . pseudo-events and 'spectacles' " (Jameson, 1991:18). When daily life increasingly consists of abstract "representation consumption," people's relationships to history, to society, to community must, inevitably, be transformed. [41]

Structuralist and poststructuralist theories of the subject have taught us that human subjectivity is constructed in the organization of social practices. What becomes of subjectivity, of desires, meanings, identities, when daily life is organized along postmodern lines? Theorists are relatively united in describing a subjectivity that is increasingly fragmented, distracted, attuned more to surface sensation and stimulation than to deeper feelings and meanings.[42]

Theorists of postmodernity agree that politics has also been fundamentally altered, but they have diametrically opposed views as to *how*. Here, David Harvey's concern about "a mine-field of conflicting notions" is indeed apt. Some theorists claim that postmodernity is liberatory. To break with modernity means, fundamentally, to break with the domination of white, male, bourgeois, Eurocentric discourses, to open up cultural space for discourses by heretofore voiceless others and make possible the celebration of multiplicity and difference. Others have a considerably more pessimistic interpretation. To them, postmodernity means that people are radically apolitical, ignorant of political facts, indifferent to public matters, unwilling to participate. Postmodern people are not able or willing to be citizens; participatory democracy, in the full sense of that term, is no longer possible. What politics is *left* is transformed into theater, into spectacle, into production and consumption of superficial political imagery. Taking place in and through the media, particularly television, political discourse becomes ever more shallow, superficial, trivial.[43]

Problems of Political Communication under Postmodern Conditions

The first, or "utopian," interpretation of postmodernity's political implications is certainly appealing. The second, or "co-optative," interpretation is, I believe, more relevant to the task at hand. It allows us to think about the conditions that make meaningful political communication and, therefore, the process of issue creation so problematic.

Production of Political Messages. In contemporary society, political discourse takes place increasingly in and through the same media circuits that provide commodified ways to fill leisure time and satisfy desires. The mass media are not neutral vehicles for communication, for discursive circulation. Stories must meet quite exacting requirements for both form and content. On television, specifically, the colorful image works better than the crafted word, and talk is best kept brief and simple. Striving for action and color, working within severe time constraints, television news reports shy away from complex explanation; social issues are, instead, dramatized and personalized.

Consumption of Political Messages. Here, critiques of postmodernity point to two kinds of problems: first, the general qualities of postmodern subjectivity; second, people's indifference to public matters. I have alluded to the first of these above. The postmodern subject is depicted as distracted, as addicted to the fleeting and fragmentary sensation, as resistant to meanings. Baudrillard says the masses refuse "the sublime imperative of meaning" (1983:7); they are a sink for all attempts to energize them.

> The masses scandalously resist this imperative of rational communication. They are given meaning: they want spectacle. . . . they idolize the play of signs and stereotypes, they idolize any content so long as it resolves itself into a spectacular sequence [p. 10]. The political has long been the agent of nothing but spectacle on the screen of private life. Digested as a form of entertainment, half-sports, half-games . . . at once both fascinating and ludicrous [pp. 37-38]. (ibid.)

Further, theories of postmodernity share with many other perspectives the view that people experience a radical split between public and private realms. To the average person, the realm of the "public," of national political events, seems distant, insignificant in comparison to the obvious importance and salience of the repetitive, mundane concerns of everyday life.

> To hear or read the news is to live intermittently in a world one does not touch in daily life. . . . Most experiences that make life joyful, poignant, boring, or worrisome are not part of the news: the grounds for personal concern, frustration, . . . the conditions that matter at work, at home . . . the events people touch, as distinct from those that are "reported." (Edelman, 1988:35)

Naturally enough, people are not interested in the "larger issues" and their knowledge of that world is, at best, hazy.

Both casual observation and social scientific studies confirm that these depictions are to a great degree justified. Signs of ignorance and disinterest are everywhere. Study after study shows that Americans know little about most public matters. Millions say they follow the news, but research finds that they do so in a superficial and evanescent manner. The steady decline in the percentage of people who vote shows that many people refuse to participate even superficially in the political process. The Markle Commission on the Media and the Electorate recently reported an "astonishing" level of ignorance about, and a "widespread, glacial indifference" toward even presidential campaigns.[44]

How can political communications proceed if these descriptions of mass political subjectivity are true, even if they are true only in tendency? What can politicians and issue makers do to overcome people's indifference? How can they capture people's attention and influence their attitudes?

A Solution: Production of Political Icons

Politicians and issue makers are sharp students of practical communications who know full well that electronic media constitute the dominant vehicle for political communication. They have learned to reduce the message to the dramatic visual and the sound bite. They also act as if postmodern theorists' most pessimistic interpretations of the state of the masses are, in fact, correct. Their solution to the requirements-of-the-medium problem and their antidote to the indifferent, distracted audience problem is the same: make political messages ever more simple, vivid, colorful, repetitive.

I would like to suggest at this point that the semiological concept of the "icon" can be appropriated to capture and summarize the quality of this change in political communications. We are most familiar with "icon" as religious representation.[45] The American semiologist C. S. Peirce appropriated the term in order to capture the uniqueness of one type of sign, "a sign which refers to the object that it denotes merely by virtue of characters of its own. . . . Anything whatever . . . is an icon of anything in so far as it is like that thing and used as a sign of it" (quoted in Sturrock, 1986:84). Sturrock's clarifying commentary is that "there is thus a picture-element in an *icon*, . . . The relation of expression to content is one of physical similarity" (ibid.:85).

Recently, as cultural criticism has come under the influence of semiological and poststructuralist theories, we have seen the term *icon*, applied to pop culture figures (Elvis, Marilyn Monroe, and the like) who are no longer just people but have been metamorphosed into fixed, highly stylized, mythical figures in the popular imagination. In fact, by 1992, journalists had seized upon the term with such enthusiasm that it was being applied too often and too loosely, to *any* actor or singer whose performing persona self-consciously evoked some stock, stereotyped cultural imagery or identity.

Still, the term has yet to be applied consistently to political phenomena or political communication. The preferred term is still *political symbol*. However, to the degree that political messages are carried by images rather than words, so that meaning or signification takes place more through nonverbal spectacle than through narrative, the term *political icon* is a more appropriate expression for what is occurring than is the more traditional term *symbol*.[46]

Lash argues that there has been a general shift in society's regime of signification, from words to images, from the discursive to the figural, from narrative to spectacle. The notion of the icon suggests that political communication has been transformed (or, if you like, degraded), exactly as Lash says. Political communication increasingly relies on the production and display of political icons rather than symbols, iconography rather than rhetoric, both because the means of communication require it stylistically and because it is assumed that displays of spectacular images are the only way to break through the indifference of the intended audience.[47]

Ambiguity in Postmodern Issue Creation: How Are Icons Received?

These concepts seem to me a fruitful way to theorize the events that made hazardous waste an issue. More generally, notions of icon production and display may be seen as an updating of the theories of social problem construction and issue creation. In the transition from modern to postmodern forms of political practice, the texture of the process of "social problems construction" is qualitatively transformed as claimsmaking rhetoric increasingly takes the form of iconography.

But issues are made not when certain messages are disseminated; issues are made when messages are believed and, better still, acted upon. We have to move from message sent to message received. What effects do icons have on the postmodern subject? What is the contemporary form or texture of mass attention? How do people receive, assimilate, and store/retain iconic political messages? Here, theory suggests two contradictory effects. Iconic communications can produce significant attitudinal changes, but those changes prove shallow and evanescent.

The production and presentation of icons—that is, of issue stories that feature repetitive, highly stereotyped, frightening imagery—speeds up the issue-creation process and makes that process take quite spectacular form, producing the kind of extreme perception of dread described by Slovic and other decision theorists. Addicted to the consumption of superficial imagery, habituated to a state of distraction, deaf to complexity and subtlety, the news consumer watches, hears, or reads news stories in a way that preserves, even enhances, their iconic quality: the strong visual and emotional components dominate; attitude formation takes place without much need for detail in the cognitive component.

Researchers have documented how media coverage of an issue produces an immediate, rapid increase in expressions of concern about that issue in the polls. This effect has been documented for nuclear power, the "environment" generally, energy, and inflation.[48] The change in people's attitudes toward hazardous waste described earlier merely confirms this oft-demonstrated effect.

As public opinion crystallizes around the highly charged images and names, perception of risk can easily outstrip officially and scientifically sanctioned estimates of the threat, in accord with the worst nightmares of Chauncey Starr and Aaron Wildavsky.[49] The postmodern *form* appears to give rise to political *effects* quite the opposite of the mass indifference and distraction depicted by pundits and theorists.

Environmental advocates certainly seem to believe in the attitude-transforming power of icons. They have mastered the art form of icon making and have used it to excellent effect, creating and successfully throwing into discursive circulation both *images*—the Baby Harp Seal, the Fifty-five-Gallon Drum, the Whale, the Cooling

Tower—and *names* that have come to metonymically connote whole fields of complex issues, such as acid rain, greenhouse effect, toxic waste, nuclear power.

Theory suggests, however, that postmodern conditions also undermine the significance or meaningfulness of such attitude formation. In a society where episodic attention is the norm, issue importance can evaporate as quickly as it forms, and nothing guarantees that even widespread political discourses will have staying power. One tunes in and consumes some imagery about environmental destruction, say Bhopal, or the *Exxon Valdez*. One cares, one can even care a great deal, during brief bursts of attention given over to monitoring national and world events. However, if such stories do not connect with something in people's immediate life sphere, if there is no material, experiential basis for the opinion formed, one will believe that a problem, an issue, exists without being more than momentarily upset about it. Absent a real connection to one's own immediate interests, concern will fade and the story will be forgotten when the kaleidoscopic flux of news moves on to other things.

Researchers have, in fact, confirmed that media-driven attitude change is unstable and short-lived. The national news may have a massive audience, but, as one study showed, even the most avid news consumer "retained . . . little more than a hazy familiarity" (Jones, 1990) with what he or she saw. Media coverage does produce an increase in expressions of concern about an issue, but it has been shown repeatedly that those expressions of concern fade just as quickly when coverage wanes.[50]

It would be easy to conclude that this kind of attitude formation is essentially hollow and politically not meaningful.[51] But that conclusion would ignore the complex, ambiguous, elusive way that perceptions and beliefs persist in people's minds. Even if media messages are consumed in a low-salience, semidistracted state, and even if focused awareness, concern, distress, or anger recedes as coverage turns elsewhere, a trace is left, a certain perceptual/attitudinal pattern is reinforced. Somewhere it is noted that uncontrolled economic activity for profit has again caused catastrophe, that neither industry nor government can be trusted to keep you safe. This belief complex, the personal trace of the discourse of modern environmentalism, will not necessarily be immediately salient in many people's lives. It may not affect their votes in elections or their other political activities. They may not report it to a pollster. It will seem absent, not there.

We must leave open the possibility, however, that attitudes and beliefs persist in some form, even if only preconsciously, that they are simultaneously absent and present, both there and not there. It is possible, then, that the right stimulus will ignite an experiential connection and vividly bring those latent attitudes alive.

Consequences

Passage of the Superfund Law

As one would expect, the sudden emergence of "toxic waste" as an issue led immediately to efforts to pass new legislation.[52] Speaking purely practically, Love Canal revealed an important gap in federal policy. The government had neither the statutory authority nor the fiscal mechanisms needed to undertake remedial cleanup activities. Under then current law, the government first had to identify the parties responsible for contamination, itself a long and difficult process, then negotiate with the responsible parties or, more likely, sue to get them to undertake remedial action. If responsible parties were found, the size of the potential liabilities—remember, Love Canal had shown that the costs of cleanup, temporary relocation or permanent buyout of victims' homes, health evaluations, monitoring, and treatment could run as high as a hundred million dollars for a single site[53]—made it inevitable that there would be protracted negotiations and/or litigation and thus long delays before actual cleanup activity could begin. If responsible parties could not be identified, there would also be delays as federal, state, and local agencies tried to hammer out ad hoc, case-by-case funding arrangements.

Of course, action on a Superfund law was not just a straightforward instance of problem solving, of lawmakers responding promptly to a newly recognized social need. The surge of media attention had made hazardous waste the nation's Most Important Environmental Problem. This was one of those quintessential "time to make a new law" moments so characteristic of the American legislative process.

The basic elements of a solution seemed clear enough. Statutory authority was needed to begin cleanups without first having to identify responsible parties, and a working fund had to be set up to pay up front for cleanup activity. This much was agreed to by practically everyone—the administration, Democrats and Republicans, liberals and conservatives, and the chemical manufacturers' lobby, the Chemical Manufacturers Association (CMA).[54] But consensus on this minimum left much to be decided and much room for controversy.

Liberals proposed a large fund, at one point more than $4 billion, paid for mostly by fees to be collected from the petrochemical industry. In addition, they proposed that the handling of hazardous industrial wastes be subject to the standards of both "strict" and "joint and several" liability.[55] This higher standard of liability would, they argued, help the government replenish the Superfund and would promote a higher standard of care. Congressional liberals also wanted the fund to provide at least partial compensation for victims' medical expenses and for lost income, and they proposed sweeping tort law reforms that would make it easier for victims to win court cases against polluters.[56]

The conservative position was that leaking orphan dump sites had to be cleaned up, but that Love Canal-type incidents were "aberrant rather than pandemic. . . . [Viewing] the by-products of industrial activity as inherently malign—a blight encapsulated in thousands of ticking time bombs capable of releasing of a torrent of poisons, toxins, carcinogens, fires, and explosions on an unwary public at any moment . . . is pure fiction" (dissenting views of Stockman and Loeffler, in Senate, 1983, vol. 2:101). Conservatives argued that a small federal program was enough: a modest fund, perhaps $600 million, paid for by general revenues or a tail-end tax on all wastes. Everything else, especially the statutory revolutionization of legal doctrines envisioned by liberals, was overreaching and regulatory overkill.[57]

It would seem, at first glance, that the liberals held all the cards in this situation, especially following that extremely photogenic explosion and fire at Chemical Control Corporation on Earth Day, April 22, 1980, and the second burst of media stories about Love Canal, the chromosome study, and the hostage-taking episode in May 1980. The issue was hot. Polls showed much public concern. All but the most extreme right fringe in Congress professed to wanting to pass some version of a Superfund law. The momentum for legislating seemed overwhelming. Something would surely pass. But how much?

Conservatives felt compelled to keep proclaiming their desire to pass a bill.[58] At the same time, they used whatever leverage and maneuvering opportunities they had to make the liberals abandon their most powerful and innovative proposals. Likewise, the chemical lobby professed strong support for legislation while its vehement opposition to the feedstock funding scheme caused much delay.

Liberals, for their part, tried to hurry things along with warnings that those who stood in the way of something the public so desperately wanted would pay a heavy price:

> [Even opponents of this bill] must be aware of this, the tremendous and volatile nature of this issue with regard to the population. . . . There was a 20-minute segment, as I understand it, on a television show called "Speak Up America," last Friday . . . It was about this very issue. . . . Just from that one program, that one 20-minute segment, came 130 letters from 28 States. . . . I do not need to belabor the point because we are all getting this kind of mail. I think that is why we will not have more than 20 or 25 votes against this bill on final passage. (Representative Moffett, during floor debate on H.R.7020, in Senate, 1983, vol. 2:263)

As things developed, as time slipped by and the 1980 elections loomed, the conservatives' tactics proved successful. The liberals had to yield on issue after issue to get the bill out of committee. By the time a bill finally reached the floor of the Senate on November 20, 1980, the political landscape had totally changed. The election had been a landslide, a political KO. The bill's supporters had to face an unpleasant choice: make significant concessions to assure passage during the lame-duck session or risk getting something even weaker when, the next year, Republicans would control both the White House and the Senate.

Even in their moment of triumph, Republicans still wanted to be seen as supporting passage of a Superfund law, but they were now also in control of the political moment and could impose severe limits on what would be passed. Marathon negotiations produced a workable, though fragile, compromise.[59] The Republicans accepted a fund of $1.6 billion, raised mostly on feedstock fees. In exchange, explicit references to "strict" and "joint and several" liability were deleted (although they were later read back into the act by judicial interpretation), and all language pertaining to medical or wage-loss compensation for victims and pertaining to a federal cause of action and other tort reforms were dropped. President Carter signed the bill on December 11, just one month before leaving office.

From Mass Perception to Mass Action

Issue formation had one other consequence of note. The period of most intense media coverage was followed by a surge of social movement activity. Were one to accept the most pessimistic, "co-opted," reading of the meaning of postmodernism, such a development would seem unlikely. An indifferent public habituated to episodic, fragmented image consumption and politicians whose behavior is dominated by the semiological and fiscal requirements of video campaigning make for a politics that is simultaneously spectacular and superficial. What do beliefs and attitudes mean if people's subjectivity tends toward superficial, distracted experiencing rather than contemplation and construction of meaning, if citizens' political phenomenology is dominated by a sense of indifference? If that is an accurate description, it seems unlikely that attitude changes would produce real political action. Postmodern people hardly seem good candidates for vibrant, grass-roots, popular democratic action.

But that depiction of social conditions is one-sided, overstated, and therefore misleading. The most cursory examination of contemporary events suggests that it describes only part of what is happening politically in American society. Indifference to public matters may be endemic, but it is far from total and uniform. People vary widely in their levels of attention and concern about issues. As a matter of everyday, habitual practice, most people may shun the larger sphere of events outside of their immediate, everyday experience; yet, there are still circumstances in which citizens find it necessary or desirable to become politically involved. In my survey of television news in 1989, for example, there were reports of protests about offshore oil drilling, commercial nuclear power plants, nuclear waste disposal, federal nuclear defense facilities, hazardous industrial wastes, pesticides, and logging of old-growth forests and rain forests. Mass ignorance and indifference coexists with the continued presence of literally thousands of voluntary associations, ranging from many small, local, single-issue groups to large, permanent, professionally run organizations with millions of members. Even if postmodern forms now saturate contemporary politics, they have not totally and decisively transformed every aspect of political practice; earlier forms, traditional movement forms and strategies, for example, persist.

Furthermore, as I suggested above, even if attitude formation takes a postmodern, semidistracted form and new attitudes seem to fade as soon as media coverage declines, those attitudes can, in the right circumstances, come alive. That, in fact, is what happened in the case of hazardous waste. The way "toxic waste" appeared on TV and in the popular press; Love Canal residents' feelings of fear and anger, their sense of injustice, helplessness, and endless nightmare; the unwillingness or inability of officials to protect people's vital interests—all of this inexorably created the perception, no matter how distractedly held, that *you wouldn't want to be in those people's shoes.*

It is this perception, perhaps more an unarticulated feeling than a conscious opinion, a preverbal sense of "not for me," that appeared in the 1980 opinion poll as a desire not to live near a hazardous waste facility and appeared subsequently as the increased likelihood that proposals to site hazardous waste facilities would arouse opposition in communities throughout the nation.

Chapter 4

The Toxics Movement: From NIMBYism to Radical Environmental Populism

In a society where episodic attention is the norm, issue importance can evaporate as quickly as it forms, and nothing guarantees that attitudes generated in an iconogenic moment will have staying power. Media-centered political communication produces a flash of worry, but, unless the problem being described connects with some genuine, immediate interest, it will be experienced as "important, but *not here.*" Concern about it, willingness even to think about it, much less significantly change one's daily activities to do something about it, fades as soon as the media move on to other things.

In the case of hazardous waste, however, the process did not get stuck in this endless postmodern loop of upset and forgetting. Local hazardous waste protest groups formed with increasing frequency. Groups began to assist each other. Soon, more permanent bonds were forged. In a mere decade, a complex social movement infrastructure was built. Ideologically, the movement grew more and more radical.

The Movement

Before 1978: Sporadic and Isolated

As reported in chapter 3, local organizing around hazardous waste issues began well before Love Canal became famous. The 1979 EPA-sponsored study of siting opposition found cases that began in 1976, some even as early as 1973. There are reports of contamination protests as early as 1970.[1]

However, the extent of such actions is impossible to determine with any confidence. Centaur Associates, hired by the EPA to study siting opposition, found "no centralized data . . . on public opposition to siting" (EPA, 1979:1). The fragmentary evidence that is available can be read two ways. On the one hand, a 1976 EPA survey of the hazardous waste management industry found that only 42 percent of TSD owners and operators felt that "public opposition was a constraint in obtaining new sites or expanding old ones" (GAO, 1978a:11), and Centaur Associates readily found siting attempts where there was no opposition or where opposition was minor and quickly dissipated. On the other hand, the consultants found it easy to assemble a sample of ninety cases of local opposition.[2] They noted, too, that participation in these actions was marked by exceptional demographic diversity, including "grandmothers and U.S. Congressmen, factory workers and university scientists, those who never graduated from high school and those with doctorates in ecology and physical sciences" (EPA, 1979:III). Overall, the evidence suggests that siting opposition was beginning but had not yet become an intractable problem.

The quality of the data on early community-based contamination organizing is, if anything, even poorer. The EPA's 1973 report to Congress lists about twenty instances of serious local contamination; by 1976, the House Commerce Committee had assembled a list of about sixty instances.[3] But it is not clear if these cases were found by local officials or were brought to officials' attention by victims protesting their plight. Examining the first wave of books on the toxics problem,[4] one finds at most several dozen contamination events that came to light because citizens protested about foul-tasting tap water, unexplained health problems, and so on.

The outstanding fact about this period was total *lack of contact* among local groups. Communities that were starting to organize acted alone. They did not seek out people in other communities who might be dealing with similar problems. Of the several dozen or so instances of local action before 1978 that are described in the literature, I found only one case where a local group reached out to activists in another community.[5] In every other instance, the organizing remained a strictly local affair. Everyone was on their own; everyone started from square one. Sue Greer of People Against Hazardous Landfill Sites (PAHLS), in Wheeler, Indiana, described how it was in an interview conducted for this volume:

Q: Did you deal with the problem yourself or did you go to outside organizations for help?

A: Really, I'm not kidding you, there were no outside organizations. . . . When we started, it was about the time that Lois Gibbs was beating the drums out in her neck of the woods but we really didn't know anything about the subject, so we really didn't know anything about the CCHW and Lois Gibbs and the Love Canal Homeowners; we just didn't know anything about those people.

Q: Were any of these larger national organizations, like EDF or EA, etc., helpful to you in terms of helping you get organized?

A: No. They did not help us get organized. We organized ourselves.

1978-1980: Actions Proliferate and Networking Begins

Contamination protests before 1978 were slow to develop because activists found it hard to convince the larger community that there was, indeed, a real problem. After the media coverage that started in 1978, people would find it easier to conceive that they, too, may be victims of toxic contamination, to construe their own situations as potentially "another Love Canal." Contamination protests that had started earlier but had floundered suddenly took off after 1978.[6] New protests sprang up. Will Collette, national organizing director for the Citizen's Clearinghouse for Hazardous Wastes, said in an interview for this volume, "The consequence of

Love Canal and Love Canal getting a tremendous amount of media play was that all of a sudden out of the woodwork hundreds of citizen groups started to form, very spontaneously, because all of a sudden Love Canal sort of gave rise to what they had been thinking and feeling."

In other communities, people now felt they *had* to defeat attempts to bring wastes into their communities. Back in 1976, less than half of facility operators had said that public opposition was a problem for them. By 1979, a GAO survey of government and industry officials found that "*virtually all* of the disposal industry officials interviewed indicated that public opposition was a major problem. . . . *Most* State officials we interviewed cited [public opposition] as the major barrier and expected public opposition to increase in the future" (GAO, 1978a:11; emphasis added). EPA's consultants, hired by administrators worried about the extent of public opposition, warned, "Public opposition to the siting of hazardous waste management facilities, particularly landfills, is a critical problem. . . . *If problems with public opposition cannot be solved, . . . the national effort to regulate hazardous waste may collapse. . . .* the prospects for successful sitings in most regions of the country are dubious at best, and grim at worst" (EPA, 1979, III, IV; emphasis added). In a 1980 letter to state governors, the EPA enunciated what would henceforth be regarded as the first truth of hazardous waste policy: "The principal difficulty . . . [in creating] new facilities . . . lies in the intense opposition of the local public to proposed (and some existing) sites" (EPA, 1980:1, 3).

Publicity and issue formation did more than just stimulate quantitative growth in the numbers of local actions; they also seem to have encouraged networking. Activists in various communities became aware of others' efforts. They made contact and began to support each other and share experiences. Will Collette said in his interview that "one hundred and fifty-some-odd citizen groups contacted Lois [Gibbs] while she was still at Love Canal." Sue Greer said that the same had happened with her some years later, when PAHLS received media attention:

> The *Today* show . . . national public attention . . . the Phil Donahue show . . . a lot of press and we did a lot of public speaking and so people learned about us rather quickly. . . . people started coming to us and saying, "Look, we see what you've done and we do not know how to do it but if you help us we are going to try it in our own community." And we did. We went from community to community helping people get started.

A review of available case histories shows the change quite clearly. Before 1978, instances of siting opposition were, with almost no exception, conducted without any contact with or help from others. After 1978, more than half the cases reported in the literature show that groups had begun to network, to bring in speakers from communities fighting the same companies, to share experiences and learn from others' tactics.[7]

After 1980: A Dynamic Social Movement

The movement then entered a qualitatively new phase. Local actions became very widespread. Formal movement infrastructure organizations developed. The movement's boundary expanded and came to encompass an increasingly diversified set of local pollution issues. The movement experienced rapid ideological evolution. Let's examine each of these developments in turn.

Quantitative Growth. Again, firm numbers are not available. No one, to my knowledge, has done a comprehensive census of local actions. It would be difficult to do so successfully. Many of these groups are short-lived. Organized in response to a specific local threat, only some of the groups stay together after the initial fight is resolved; others disband quickly once the original focus is gone.[8] Second, the boundaries are ill defined. As I will describe below, the "hazardous waste" movement has broadened into a larger "toxics" movement that now encompasses a much larger set of concerns. The core movement infrastructure organizations have the best raw data, the telephone logs and contact sheets that record requests for help from local groups, but the organizers are overworked, and keeping good records, they noted in interviews, is not a priority for them.[9] Without firm numbers, we must make do with meaningful indicators. The most persuasive indicator is that both sides agree that local organizing has reached unprecedented proportions.

The Citizen's Clearinghouse for Hazardous Wastes publishes an annual count of the grass-roots groups it has served in some way: more than 600 groups by the end of 1984, more than 1,000 by the end of 1985, 1,700 by 1986, 2,739 by 1987, 4,687 by 1988.[10] In 1988 alone, CCHW claims to have physically sent organizers to work with citizen groups in 162 cities in thirty states and the District of Columbia.[11] CCHW regional organizers told us that, at any one time, they are in phone contact with dozens of groups in their regions.[12] The two other national-level infrastructural organizations, the National Toxics Campaign (NTC) and Greenpeace, say they have worked with about 2,000 and 1,000 local groups, respectively.[13]

These numbers cannot be verified independently, thus one cannot dismiss the possibility that they may be inflated. Government and industry spokespersons presumably have no reason to exaggerate the extent of local organizing, a phenomenon they fervently wish did not exist. Waste generators, the disposal industry, consultants, state and federal officials, lawyers, and policy scientists all agreed that local organizing had become very widespread and that local opposition had become the biggest impediment to facility siting.

Leading generators stated:

Public opposition . . . has blocked the construction of new or expanded facilities, although the need for such facilities is obvious. (Chemical Manufacturers Association, n.d.:6)

Legislators and regulators have inflamed public opinion against hazardous waste facilities. Time and time again, the public is terrorized into blocking the construction of responsible treatment and/or disposal facilities. (William L. West, Republic Steel, 1985:39)

The waste disposal industry concurred:

The "siting problem" for hazardous waste facilities is the most serious single obstacle remaining to be overcome. (Richard L. Hanneman, National Solid Waste Management Association, 1982:10)

State officials and their consultants added their voices:

There have been many more failed siting attempts than successful ones. . . . The one factor these failed proposals all have in common is extensive public opposition. (survey of siting in the United States, done for the state of Massachusetts; Ryan, 1984:8)

The most formidable obstacle to Waste-to-Energy facilities is public opposition. (Cerrell Associates, consultants for the state of California, and Powell, 1984:3)

It appears that the real obstacles to hazardous waste management are the potential legal entanglements arising out of citizen opposition rather than . . . technical requirements. (James MacAvoy, Ohio Environmental Protection Agency, 1980:450)

Federal officials also expressed concern:

If we are not able to convince the American public that RCRA will ensure that disposal sites are safe and well managed, we will never acquire the needed and necessary sites for facilities and, therefore, never achieve the objectives of RCRA. . . . the siting issue is the most difficult problem facing us in the implementation of RCRA. (Thomas Jorling, EPA assistant administrator for water and waste management, in Senate, 1978:15)

Why has it been so difficult to site hazardous waste facilities? The fear of risk to public health is obviously the major impediment. The American public is traumatized by hazardous chemicals and the possibility of contracting cancer or birth defect diseases. (Alvin Alm, EPA official, 1984:2)

Other, "neutral," observers commented also:

Some people's exaggerated perceptions about risk, however, may be the most serious obstacle to successful siting of new facilities. (Keystone Center, 1980:21)

The public is opposed to any type of management. . . . The result has been gridlock and almost no new facilities. (Editor, Hazardous Waste and Hazardous Materials, 1988:ix)[14]

Some pointed out that the demographic base of opposition groups was proving to be unusually diverse:

The overriding conclusion of survey research on public attitudes toward these facilities is that opposition to the local siting of such a facility cuts across all subgroups. Regardless of socioeconomic status or residence, specific cases can be found in which even the subgroup least likely to form an opposition movement became intimately involved in the opposition struggle. (Cerrell Associates and Powell, 1984:18)

Almost the entire community . . . turns out to be environmentalists—the Kiwanis, the Rotary, the League of Women Voters, sometimes even the banks and the Chamber of Commerce. The environmental position is taken by almost everyone. (Robbins, 1982:505)

During the 1980s, the question of what to do about local opposition became *the* dominant topic in official hazardous waste policy discourse. Considerable intellectual energy was expended in the effort to figure out how to neutralize the power of community organizations. I will examine this "discourse of disempowerment" in some detail in chapter 5. For now, I note simply that the very presence of all this obsessive concern suggests that local organizing had, indeed, become ubiquitous.

Building a Formal Organizational Infrastructure. Informal networking proved very helpful. People learned they were not alone.[15] They gave and received the emotional support so necessary to keep themselves going when, inevitably, their will to keep fighting flagged. Those who had been involved longer helped those who were just getting involved.[16]

Formalizing these contacts was a natural step. Fittingly enough, the first and still most important of these formalizations, the Citizen's Clearinghouse for Hazardous Wastes, came out of the Love Canal experience. As the calls for help from other communities kept coming, Lois Gibbs, the leader of the Love Canal Homeowners' Association, decided to set up an organization that could "give other people the kind of help she wished she had gotten when she first started at Love Canal" (CCHW, 1986b:15-16). In the main, CCHW defines its mission as classic Saul Alinsky-style community organizing: "to help people help themselves. . . . 60 percent of staff time is spent on the telephone helping people think through how to start local groups. . . . Our organizing staff and field staff spend many weekends visiting communities to speak at public meetings and provide training" (CCHW, 1989:49).[17] CCHW says it has worked with more than 5,000 local groups. CCHW also sponsors regional Leadership Development Conferences (LDCs), where local leaders from different communities come together to network, share experiences, and sharpen their political skills.[18] Periodic national conventions help build the sense that all the separate local groupings CCHW works with are part of a real Movement.

State and regional groupings were also beginning to get organized. This happened in a number of ways. In some areas, several autonomously formed groups decided that local action alone was not enough to take on opponents that were operating on at least the state, if not the regional or national, level. In others, a particularly strong group decided to expand to other communities or to become a regional resource center for others. In still other places, a national-level organization, such as CCHW, was aware of groups forming near each other and brought them together, hoping they would go on to forge a more permanent regional "hub." Today, state-level coalitions are present in every region of the United States, the Deep

South, New England, the industrial Great Lakes region, the Mid-Atlantic, the rural Midwest, North Central, and the Far West.[19]

New national-level organizations developed. The National Campaign Against Toxic Hazards originated in 1984 as a coalition to strengthen the Superfund law. After 1986, now renamed the National Toxics Campaign, the organization adopted a hybrid strategy that combined national policy campaigns with local organizing. Nationally, source reduction is at the heart of everything NTC does and says. NTC also hones in on a few high-profile policy issues, opposing toxic waste incinerators, for example. At the local level, NTC says it has worked with about 2,000 organizations, offering these groups CCHW-style organizing help, technical assistance, and laboratory testing.[20]

The movement has generated its own information services. The Environmental Research Foundation in Princeton, New Jersey, for example, keeps abreast of scientific research being conducted on all aspects of the toxics problem. It tells activists how they can locate this information; it identifies the most important findings and makes them accessible in understandable, usable form, in a computerized data base, the Remote Access Chemical Hazards Electronic Library, or RACHEL, and a weekly newsletter, *RACHEL's Hazardous Waste News*. The movement is served by a number of other computer networks, such as Greenpeace's bulletin board, Environet, and by numerous newsletters.

It is important to note, too, that as it has developed, the hazardous waste/toxics movement has crossed class and race boundaries. In its previous phases, American environmentalism could plausibly be dismissed as middle-class and white.[21] With the maturation of *this*, the grass-roots or toxics phase of environmentalism, that is no longer an accurate portrayal.

The industrial facilities that produce toxics tend to be located in or near communities of the working poor and of people of color.[22] Historically, waste disposal facilities were, likewise, to be found in the poorest, demographically most heavily black or Latino communities.[23] As facility siting became more difficult in the 1980s, some policy analysts began to advocate a strategy of siting in communities that are least capable of politically resisting or most amenable to accepting some form of financial compensation in exchange for accepting the facility.[24] Recently, for example, there has been a flurry of attempts to site hazardous waste facilities on Native American land.[25] Antitoxics environmentalism, then, is an environmentalism to which working people and people of color can relate. By 1990, toxics organizing in the nation's racial and ethnic communities was perhaps the most dynamic, fastest-growing facet of the toxics movement.

In the South, black churches helped organize opposition to a proposal to dispose of PCBs in Warren County, North Carolina, in 1982. Since then, the churches (the Church of Christ and its Commission for Racial Justice), historically black colleges (such as Albany State College in Georgia), older community organizing centers (Highlander Center), and CCHW have all worked to provide technical assistance and organizing skills to low-income black communities that are dealing with waste or toxics issues.[26]

Native Americans for a Clean Environment (NACE) started in Oklahoma in 1985 as opposition to an injection well proposed by Kerr-McGee. Today, NACE, with help from CCHW, works with Native American nations who are resisting the siting of disposal facilities on their reservations.[27]

In the Southwest, the Southwest Network for Environmental and Economic Justice, a network of people from both Latino communities and Native American reservations in seven western and southwestern states, provides "leadership development and technical assistance to communities of color."[28] In California, California Rural Legal Assistance (CRLA) has expanded its work from traditional farm worker issues, such as the basic right to organize, to the question of farm workers' exposure to pesticides. With the help of CCHW, CRLA now works with minority communities resisting hazardous waste facility siting in rural California.[29]

By the end of the 1980s, then, "the hazardous waste movement" consisted of a vast, somewhat loosely articulated network of many local (often short-lived) groups, statewide and regional coalitions, and national-level organizations that provided sophisticated information services and organizing assistance.

In addition, the older, more established environmental organizations had also taken up the cause. Of these, Greenpeace took the most radical line, advocating source reduction, fighting the export of wastes to less developed nations, helping communities, CCHW-style, to oppose incinerator siting.[30] The others—the Sierra Club, the Environmental Defense Fund, the Natural Resources Defense Council, the Audubon Society, the National Wildlife Federation, and the Conservation Foundation—all, to varying degrees, incorporated toxic waste issues into their routine lobbying and litigation efforts.[31] Their presence provides the movement with more conventional means for exercising influence within the proceduralized, normalized realm of "Washington politics."[32]

Issue Expansion. The toxic by-products of industrial production are hardly the sole source of localized environmental degradation. A community can face threat from a host of other sources—the smokestack emissions of a local factory, unsafe disposal of infectious hospital wastes, toxics stored at a nearby military base.

Over the decade, local groups sprang up around many such issues. They, too, contacted the organizations of the hazardous waste movement and asked for help. Core movement organizations may have begun with single issues, often just single facilities, but, over time, as they responded to requests for help, they expanded their focus to other types of problems.

Initially, CCHW dealt almost exclusively with Love Canal-like cases of community contamination. By 1986, CCHW was helping communities fight facility siting (both hazardous and solid waste sites, landfills and incinerators, waste-to-energy facilities). It was helping other communities deal with deep well injection, military toxics, pesticides, industrial plant emissions, and transportation of wastes.[33] By 1988, CCHW literature was talking about global issues such as the greenhouse effect and ozone depletion; the Five-Year Plan of Action adopted in 1989 resolves to "broaden [the] movement [to] include *all* environmental hazards."[34]

Gary Cohen of the National Toxics Campaign said in an interview:

> Originally, the coalition [NTC] was really around the dump sites . . . around Superfund sites. And then it broadened out to incinerator sites. And then it broadened out to solid waste incinerator sites. And then it broadened out to people that are actually fighting the firms, themselves, the producers—the IBMs, the Dows, the DuPonts, . . . now it's also starting to branch out into the military toxics sites.

Sue Greer, of PAHLS, the Indiana coalition, told an almost identical story of issue expansion in a local organization:

> When we first started out we did strictly landfills. And then, as the years went on, we got into air pollution, and hazardous waste, and as the garbage problem grew and the landfill situation got worse and they started turning to incinerators and this loophole in RCRA on the burners of blenders where they can burn hazardous waste as an alternative fuel, with no permits, just an EPA ID number. . . . We got into wetlands. It was just one thing at a time. We didn't mean to take on all those issues. We did not actively ever go out and seek a new issue; it came to us.

In this way, the "hazardous waste" movement gradually became a much more broadly defined "toxics" movement.

The process of issue expansion is continuing. The PAHLS newsletter now carries stories about an even wider set of environmental issues: pesticides, radioactive waste, PCBs, military nerve gas, food irradiation, ozone depletion, electronic pollution, acid rain, global warming, oil spills, Agent Orange. The Silicon Valley Toxics Coalition, organized, originally, to fight water pollution caused by the "clean" production methods of the high-tech computer industry, now gives prominent coverage to both atmospheric ozone depletion and the problem of global warming. The Environmental Research Foundation's newsletter may still be titled *RACHEL's Hazardous Waste News*, but it now carries stories about solid, medical, and radioactive wastes, nuclear power and nuclear weapons, occupational health hazards, harmful consumer products, agricultural chemicals, air pollution, acid rain, and ozone. In an article on movement strategy, *RACHEL's* editor, Peter Montague, wrote:

> It would be beneficial to broaden the . . . concerns of our Movement . . . [to] pollution of the workplace; gross industrial air pollution of whole cities and regions; pesticide pollution and industrial pollution of our food; automobile pollution; indoor air pollution. . . . And in the next few years, grass-roots activists will need to pay serious attention to what look like bunny-hugger issues but aren't: habitat destruction, loss of wildlife species, and loss of wetlands. (1989:104)

Ideological Development: From NIMBYism toward a Radical Environmental Populism. The hazardous waste and toxics movement is often characterized by its critics in industry and government as a NIMBY, or "not in my backyard," phenomenon. Examined more closely, this catchy, seemingly simple and straightforward term invokes several related accusations about the nature of the movement and the people who are active in it.

People's oppositionism is said to be *based* on faulty, irrational perceptions. People are said to overestimate wildly the risks of hazardous wastes. They have supposedly become so phobic that they cannot distinguish between the admittedly dangerous, high-risk disposal practices of the past and the much safer, technologically advanced, tightly regulated disposal methods of today. Comments about overestimation of risk include the following:

> The American public is traumatized by hazardous chemicals. (Alvin Alm, EPA, 1984:2)

> All hazardous wastes are [assumed] life-threatening and cancer-causing. (Peggy Vince, Ohio siting official, 1982:20-21)

> Inflamed public opinion. (William West, steel executive, 1985:39)

> In some cases near-hysteria. (Richard Hanneman, National Solid Waste Management Association, 1982:9)

> Buzzwords like dioxin inflame public fear. (Editor, *Hazardous Waste and Hazardous Materials*, 1988:ix)

Comments about the inability to distinguish between safe and unsafe disposal include these:

> Publicity has focused almost exclusively on the disastrous results of improper management of hazardous wastes. The public is thereby unable or unwilling to distinguish between patently improper sites for hazardous waste disposal such as Love Canal, and properly managed disposal sites. (Centaur Associates, in Duberg et al., 1983-84:85)

> No distinction is made between illegal dump sites and legitimate TSD facilities. Rather than seeing properly operated management facilities as part of the solution, a means of preventing illegal dump sites, they are perceived as one and the same thing. (Vince, 1982:21)

> Much of the opposition . . . is attributable to the public's confusion between the old generation of poorly managed and poorly regulated facilities and the new generation of well-monitored disposal facilities. (Cerrell Associates and Powell, 1984:10)

Unfortunately, the argument goes, people's irrational, fear-driven militancy can produce terrible results. If successful, opposition to facility siting can harm economic performance. And, ironically, it can actually put communities and society in greater danger because wastes must go *somewhere*; if not to properly designed and permitted sites, then surreptitiously to municipal landfills, to midnight dumpers, the mob, down the storm drain. People do not understand the tragic paradox that the sum of many communities' fear of pollution makes it inevitable that *some* community, and society as a whole, will certainly be polluted. Those who predict dire economic consequences make statements such as these:

> Without adequate facilities, needed goods and services simply cannot be produced. (Chemical Manufacturers Association, n.d.:6)

> Hazardous waste management facilities are needed . . . to assure the smooth functioning of the many industries generating hazardous wastes as a result of providing valuable products for the United States. (Keystone Center spokesman, in Craig and Lash, 1984:100)

> Industrial leaders in Pennsylvania have voiced concern over the potential outcome of such an unfortunate situation. Some have expressed that at worst, it could lead to loss of business, loss of jobs, loss of tax base, eventually even to loss of population. (Buckingham et al., 1986:500)

> The safe disposal of hazardous and toxic substances is of enormous concern; if the problem is not solved, it will severely limit industrial growth. (Governor's Commission on Science and Technology for the State of New Jersey, 1983:18)

> We have created a shortage of facilities . . . so that the companies with this waste will pay almost any price to get rid of it. . . . it is well to remember that costs to industry come out of the national assets just as surely as tax dollars. (Editor, *Hazardous Waste and Hazardous Materials*, 1985b)

Those who argue that, ultimately, harm to the environment will be caused by opposition to waste disposal sites offer these statements:

> The simple truth of the matter is that the waste has to go somewhere—we cannot shoot 40 million tons of it off into space. If we do not establish environmentally sound disposal sites, the inevitable consequence is that the waste will wind up in our backyards anyway—but without the controls that would keep it from doing us harm. (Barbara Blum, EPA official, in Morrell and Magorian, 1982:13)

> The shortage of suitable facilities will be further aggravated by public opposition to siting new ones. Unless the opposition gives way . . . excess wastes will continue to be disposed of in dangerous ways—via midnight dumping and endless on-site storage. (League of Women Voters, 1980:8)

> Environmentalists have simply gone too far in blocking in LULUs [locally unwanted land uses]. . . . If new hazardous waste facilities cannot be sited, the waste must still go somewhere—to existing overburdened facilities, or often to organized-crime fronts, to midnight dumpers. (Popper, 1987a:2, 10)

> Ironically, but sadly, this opposition [to new facilities] may be leading to situations that could seriously threaten public health, including, for instance, illegal dumping of wastes on roadsides. (Keystone Center, in Craig and Lash, 1984:104)

The New Jersey Legislature recognized that because of the critical shortage of suitable sites to properly handle hazardous waste, there was grave potential that future waste disposal sites would be selected "on an indiscriminant and illegal basis." (Frank J. Dodd, New Jersey siting official, 1986:424)

[Capacity shortage in Massachusetts] has created an enormous incentive for what is called midnight dumping. (Massachusetts environmental official, in House, 1982b:21)

If generators have no place to get their wastes treated, they have two options available: increased long-term storage or environmentally unsound disposal practices (midnight dumping, for example). Neither is desirable, but both are on the increase. But what does a small business man do when he runs out of storage space and no one will take his wastes? Go out of business? (Suellen Pirages, National Solid Waste Management Association, 1987:34)

People are said to be so narrowly self-interested that they are either unaware or unconcerned about the dire economic and even environmental consequences of their refusal. And, if local activists do understand that, they are, even worse, utterly selfish. Such people like the benefits of industrial production, but they don't want to bear the costs:

Citizens groups . . . fail also to accept . . . the need for solutions. "Put it in Texas," is a convenient argument for local use (unless you're in Texas), but it merely passes the buck and denies the fact that those who benefit from technological advancements must also share the burden of responsible management of its by-products. (Vince, 1982:20)[35]

I will consider the charges of harmful economic and environmental impacts later, in chapter 7. Here, I wish to limit my discussion to the question of subjective intention or, to put it another way, the consciousness (as opposed to the actual consequences of the actions) of the movement and its participants.

Movement leaders readily admit that they started out years ago with a limited, NIMBY consciousness. Sue Greer, for example, said, "In the beginning, we were as guilty as the rest of them, . . . we wanted that thing [the hazardous waste dump down the street] to get the hell out of there. And, I mean we thought, 'get it out of here and take it anywhere but don't have it here' " (interview).

But that is no longer true today. I have already described how the movement expanded from hazardous waste to a host of other toxics issues. That expansion was accompanied by an increasingly comprehensive, totalizing critique of modern economic production and forms of political power.

When the problem was conceived of as "our contaminated community," the cause was "that landfill" or "that careless chemical firm." When the problem was conceived of as hazardous industrial waste, generally, the cause of the problem was Waste Management, Inc., Browning-Ferris, the disposal industry, polluting firms, do-nothing state and federal officials. As the movement grew and addressed an ever larger set of problems, the cause came to be defined very broadly, in terms of a whole system of technology and chemical production, driven by profit, unchecked by a government that serves private wealth rather than public interest:

> The toxics crisis is not simply a hazardous waste crisis. It is a crisis involving a four-decade explosion of technologies which pose public health and environmental threats unprecedented in nature and scale. . . . petrochemicals and synthetics . . . high-chemical agriculture . . . the nuclear industry for power and weapons . . . microelectronics. (National Toxics Campaign, n.d.:2)

> We are endangered because the polluters, in their quest for profit, have exhibited a callous disregard for the lives and health of people. The corporate polluters have used their enormous power to . . . weaken regulation, and to avoid paying damages for injuries they have caused. . . . Government has failed to protect the American people. The regulatory program is ineffectual. (ibid.:1)

> America's largest financial interests and the influence in government their money can buy. (CCHW, 1988:n.p.)

The movement has not settled on a single, clear political ideology. Rather, one finds an untroubled eclecticism, a coexistence of multiple political symbol systems that have little in common except that they can be mobilized to legitimate a position of radical critique and activism.

Some in the movement legitimate their radicalism by appropriating the symbols of the American Revolution, the Declaration of Independence and its triad of inalienable rights, "life, liberty and the pursuit of happiness," and the Bill of Rights, and by emphasizing the radical, social movement aspects of the Revolution, such as the Boston Tea Party.[36] Others tend, instead, to justify their environmental radicalism through an ecological reinterpretation of the Bible and of Christ's teachings.[37] A third, different referent is activated when movement newsletters reproduce Chief Seattle's purported 1854 letter to President Franklin Pierce, a document that expresses traditional Native American respect for and harmony with nature.[38] The fact that this letter is almost certainly a fabrication doesn't matter; even if inauthentic, it allows an identification with an imaginary other, a nonmodern, premodern, non-Western perspective from which point of view contemporary society's treatment of Nature appears an abomination.

Beyond such specific ideological referents, the clearest and most consistent ideology of the movement is a rather traditional, rough-hewn but, for all that, quite durable and serviceable populism that depicts American history as, most centrally, a struggle of the small people against big government and big business. It has always been "the people" against the privilege and power of dominant, exploiting, selfish, and uncaring elites. Environmental damage today is merely the latest in the long list of injustices that have been caused by "greedy corporate polluters and their friends in government" (CCHW, 1986b:4). Similar sentiments are expressed by other movement organizations and by key movement leaders:

Corporate polluters . . . their quest for profit [and] callous disregard for the lives and health of people. . . . Government [that] has failed to protect the American people. . . . *"Who pays and who profits?* . . . the polluters profit while the rest of us pay!"* (National Toxics Campaign, n.d.:1, 4)

Government bureaucrats who have lost their connection to the people they serve; . . . corporate decision-makers who have lost their connection to the people affected by their decisions. (Montague, 1989:104)

Government and large corporations, . . . greed and power. (Lois Gibbs, "Together, We Can Win Justice," in CCHW, 1989)

Sometimes, the grass-roots environmental movement is depicted explicitly as squarely in the tradition of American populism: "It's an old-fashioned Movement that addresses old-fashioned American values of neighbor helping neighbor, of grassroots democracy where the people lead and the leaders follow" (CCHW, 1986b:41). It seems appropriate, today, to describe the movement's ideological position as a *radical environmental populism.* This phrase *situates* the movement in a larger history of American radicalism while it *distinguishes* the movement both from earlier forms of populism and from other tendencies in contemporary environmentalism.

Totalizing concepts of root causes imply radical cures. The movement is not explicitly socialist or anticapitalist, but the reforms it advocates would, if adopted, amount to a fundamental restructuring of the current relationship between economy and society.

Every organization in the movement agrees that pollution prevention through source reduction is the key to really solving society's toxics crisis. CCHW's Plan of Action, for example, calls for mandatory recycling, mandatory source reduction, and outright bans on the production of harmful toxic materials. NTC calls for "toxics *prevention* as the primary focus of environmental protection" (n.d.:5). Source reduction is one of Greenpeace's three principal goals.[39] Barry Commoner, the most famous environmentalist now associated with the movement, makes explicit the radical nature of the demand for systemic source reduction by saying that it is tantamount to a demand for the "social governance of the means of production" (1989:12).

The movement's leading groups envision a radical democratization of politics. CCHW says that social governance of production, if it ever happens, should not mean governance by officials and bureaucrats, but "direct citizen representation in *all* decision-making" (1989:57-60). NTC calls for "measures at every level of government to empower citizens and communities in their efforts to protect public health and community well-being" (n.d.:5-6).

Ultimately, the movement's ideological radicalization leads it to make common cause with other social justice movements and to embrace a much broader progressive agenda. I will discuss that development later, in chapter 8.

The toxics movement can no longer be characterized as narrow, selfish, shortsighted—quite the opposite. Even if, tactically, the movement still mostly takes the form of local actions against single local targets, those local actions are informed more and more by explicit, long-term goals of radical social change and by visions of a just society.

It is true that people who are just becoming active today still tend to start from a narrow, NIMBY position. But when they contact the movement's infrastructure to ask for help, they not only get help, they get a full dose of the movement's radical analysis. And they are exhorted to stay involved, become part of the movement, help others.[40]

Charges of NIMBYism may have been justified a decade ago; they are no longer fair or accurate today. I suspect, however, that for government and industry there is little comfort in the fact that NIMBYism has matured into a quite radical environmental populism.

When Movement Follows upon the Iconogenic Moment

As I write this, news coverage of the 1992 presidential contest is dominated by a debate between the vice president and a television sitcom character. Just a few years ago, Hollywood's ability to combine cartoons and live actors in *Who Framed Roger Rabbit?* was hailed as a technical tour de force. Today, this real-life conflation of the imaginary and the real hardly stirs people to think it bizarre. (A *New York Times* writer did comment, "If there was still a distinction between politics and entertainment in America before Monday night's episode of 'Murphy Brown,' there did not seem to be one afterward"; Kolbert, 1992.) The conduct of this election has done little to refute the claim that postmodern processes are an important feature of contemporary politics.

On the other hand, the election also shows that postmodernism has not totally taken over the whole of the political process or the production of forms of consciousness. Ultimately, bad economic conditions undermined any chance of the incumbent's reelection. Thus, it seems that no amount of image manipulation can keep people, finally, from voting their pocketbook, that is to say, something like their "real," material interests.

I argued in the last chapter that postmodernists' claim that society has entered into an *absolutely* new period is an overstatement. Cogent as the descriptions of postmodernity are, they describe only part of what is happening in American society. Earlier forms of social practice persist; contemporary politics consists of some amalgam of postmodern, modern, and even, at times, premodern forms.

That coexistence of different forms of political action opens up the possibility that an issue that first takes postmodern form may subsequently trigger other forms of action. If the media images connect with personal experience, with what are perceived as immediate, compelling interests, an icon can become the perceptual basis for a more traditional politics of the social movement.

That shift from postmodern issue creation to more traditional social action is what happened in the case of toxic waste. The attitudes and beliefs generated by media images of Love Canal did not just persist as a distanced, somewhat abstract "concern" about another "problem." When the possibility arose that something similar might happen to them, or, indeed, that it might already have happened, people proved more than willing to take the time and trouble to get involved.

Sociological research has identified a number of conditions or factors that favor the emergence of social movement activity.[41] With the help of this literature, we can explore what it was about the two types of local hazardous waste situations— communities already contaminated and communities proposed for new facility siting—that made it easy, in fact subjectively almost imperative, for people to move from perception to action.

Conditions That Favor the Emergence of Social Movements

Threat as Motivation. The social movement literature emphasizes that discontent alone does not cause social movement; some discontent is *always* present in society. Movements begin when conditions worsen and grievances are more deeply felt. They are often triggered by "suddenly imposed grievances" (McAdam et al., 1988:706). The attitudinal effect of what I have been calling *icon formation* was that any sudden discovery that one was living, or might be made to live, in the proximity of toxic waste would be experienced as a serious worsening of one's condition and as a deeply felt grievance.

The Contamination Experience. Contamination episodes are the stuff of nightmare.[42] Typically, the hazard is not directly visible to the victims. It is known to be there or purported to be there, but its extent, its intensity, the actual level of threat it presents, cannot be gauged.

> Most contaminants are environmentally and medically invisible. . . . not detectable by the senses . . . there are no definite answers . . . it is impossible for [people] to determine with any degree of certainty if they are in danger. . . . Behavior sufficient to avoid the hazard is not possible when individuals cannot estimate the extent of potential injury or damage. . . . The more invisible to hazard agent, the more difficult it is for people to receive feedback on just how effectively they are coping. (Kroll-Smith and Couch, 1991:7-8)

Worse, even if contamination episodes have a clear, identifiable starting point, an explosion or a spill, they persist for long periods and typically don't have a definite end-point after which the threat is over, people can relax, and normal life can resume.

Threat that is omnipresent, but invisible; pervasive uncertainty; no end in sight—these conditions produce extreme emotional reactions: "intense fear . . . chronic anxiety . . . demoralization . . . feelings of helplessness" (ibid.:6, 7, 8). Unable to bear the pervasive sense of uncertainty, people tend to develop firm, subjectively certain beliefs about the "true" situation.[43] They have no doubts that their bodies and their loved ones' bodies have been violated, polluted, poisoned.

In his famous history, *The Making of the English Working Class*, Edward Thompson notes that nascent capitalism violated a deeply rooted sense of moral economy left over from feudalism, and he shows how this sense of violation contributed to class formation. Capitalism has its own moral economy, some of it derived from the idealized ethics of markets, some of it derived from the principles of the bourgeois/democratic states that rest their legitimacy on due process and on the fairness and responsiveness of officials. This sense of capitalism's moral economy may be held in an inarticulate, largely implicit fashion, but it is deeply felt and perceived violations give rise to powerful feelings that norms of fairness have been violated.[44] People who find themselves the innocent victims of toxic contamination feel that, in addition to the material violation of their bodies, their moral-economic sense of fairness or justice has been grievously violated. That feeling is only intensified as the victims find that officials are unable or unwilling to remedy their plight.[45]

The very invisibility of the purported hazard, however, used to make it difficult for people to construe themselves as contamination victims. Media coverage of protests in contaminated communities changed that. As decision theorists' analysis of "availability" suggests, media coverage made it more likely that people would construe vague and ambiguous experiences as evidence that they have been contaminated.[46]

Siting Opposition. People had not always opposed the siting of hazardous waste facilities in their communities. The EPA survey from 1973 (chapter 2, Table 1) suggests that, at that time, people had not yet learned to fear such facilities. By 1979-80, all the evidence confirms that attitudes had changed and siting opposition had become widespread.[47] Given what they saw on TV and read in the papers, people had plenty of reasons to oppose siting.

Recall, first, Popper's analysis of locally unwanted land uses, that is, facilities and infrastructural construction projects that provide benefits for society at large but concentrate all the costs and disamenities on the host community. People living near proposed LULUs, Popper tells us, are rational to resist having such costs imposed on them. The iconography of hazardous waste fostered the belief that hazardous waste facilities are ticking toxic time bombs, that people unfortunate enough to live near them risked serious health problems and financial ruin. The icon, in effect, tagged hazardous waste facilities as the worst imaginable LULU.

The Centaur Associates/EPA study of opposition documented the extraordinarily wide spectrum of interests that people felt would be at risk were a hazardous waste facility constructed in their community. Concerns for health and safety were raised most insistently. Such concerns included fear of catastrophic events such as fires, explosions, and toxic spills; fear of insidious/chronic pollution, especially groundwater contamination; and fear of what Centaur Associates called "political wastes," substances that had been extensively publicized, such as PCBs and dioxin. A second cluster of interests, mentioned less often but still important, involved nuisance and undesirable life-style impacts. Nuisance included concern about dust, noise, odors, traffic, and rodents. Life-style issues included both general interference with local life-style and more specific concerns with visual impact.

Fear of economic impacts constituted a third cluster of concerns. These ranged from fear of adverse impacts on individual property values to more general threats to the community's dominant economic activities, such as tourism and agriculture. Proposed TSD facilities might, it was feared, interfere with more desirable community goals and land uses, such as planned industrial development, housing, or hoped-for development of recreational space.

I noted above that contamination violates deeply held standards of moral economy. Perceived violation of moral economy is also important in siting opposition. In the Centaur case studies, moral-economic threats appeared in several guises. Residents felt it was wrong that their community would be made to take other communities' wastes. In some communities, residents were upset to discover that the nature of the proposed facility had been misrepresented to them. Citizens often felt that regulators, who, after all, ought to be serving them and assuring their safety, seemed unresponsive to citizens' concerns. Citizens also felt that their nominal right to "participate" was phony, that powerful economic and political actors elsewhere would ultimately do whatever they had planned to do anyway, regardless of local sentiment.

Stigma—the threatened loss of social status, the symbolic soiling of the community and its image—was also a powerful motivator. Being a dump, being dumped on, being forced to live with others' waste products—undoubtedly, we are dealing here with a *double* stigma. The presence of a landfill or other waste treatment facility signifies the *social* stigma that the community (and by implication its residents) is at the bottom of society's class and status hierarchy. This overt, consciously felt stigmatizing signification is underpinned and emotionally fueled by a psychodynamic subtext of fecal imagery. "Waste," "dump," "being dumped on" all carry with them preconscious associations with feces, being unclean, and so on that surely contribute to the sense of repulsion at the thought that one will have to live with someone else's (industrial) shit.[48]

Capacity. Even if people have the *motivation* to act, the research on social movements finds that their *capacity* actually to do so will be affected by a series of conditions. Mobilization is more likely during periods of economic prosperity. Mobilizations are more likely, also, when the political system is in crisis or is, for whatever reason, momentarily more vulnerable and therefore responsive to protest. Protests are more easily organized if potential members have some preexisting basis for associating, such as the community or the workplace, and still more easily if potential members have some prior organization. People are more likely to join if the anticipated benefits of participation outweigh the perceived costs. In the case of the hazardous waste movement, some of these expectations are confirmed and others are not.

That in these situations the benefits of activism far outweigh the costs seems obvious enough. The contamination experience is so terrible, such a physical, moral-economic, and therefore psychological violation, that once one is convinced that one is a victim, the benefits of doing something would seem far more attractive than resignedly bearing the costs of doing nothing. Research on reactions to contamination does suggest that when faced with that situation most people's impulse has been flight, not fight.[49] However, seeing on television that some communities were dealing with their plight by protesting and seeing that their protests were getting them at least some of what they demanded—medical tests, temporary relocation, even home buyouts—increased the likelihood that others would choose that more militant response.

Research shows that most communities slated to host a new hazardous waste facility feel that there is little or nothing to gain and potentially much to lose if they accept the siting. Communities have been known to accept, even to fight to retain, a polluting industrial plant if that plant provides the community significant benefits. But a hazardous waste facility promises few, if any, benefits for either the community or the individual—some jobs, perhaps, and some contribution to local taxes.[50] In contrast, having a facility nearby carries with it high perceived costs, as described above.

The only thing that could possibly have altered people's evaluation of costs versus benefits would have been increased trust, increased belief that regulators are able and willing to do what it takes to make these facilities safe. But that failed to happen. Problems with early RCRA implementation were widely publicized. There were reports of widespread midnight dumping and organized crime involvement in hazardous waste hauling and disposal. After 1980, both RCRA and Superfund suffered as the Reagan administration vigorously pursued a policy of deregulation. The "Sewergate" scandal at the EPA, in 1983, forced some improvements, but neither RCRA nor Superfund was ever implemented by that administration well enough to deserve the public's confidence.[51] Given unremitting implementation failure, it seemed only a matter of time until the community with a facility became, inevitably, a contaminated community. One could protect oneself only by keeping the facility out of the community in the first place.[52]

Few, if any, benefits are forthcoming if the facility gets built, and potentially catastrophic costs may result. The costs of getting involved to prevent the siting, some time and some effort to go to a few meetings or to show up and demonstrate at a public hearing, seem low in light of the potential benefits of successfully warding off the anticipated threat.

The social movements literature says that geographic proximity and concentration of potential recruits facilitates mobilization, as does a history of prior organization. The first condition, proximity and concentration, is met automatically in these cases; the locus of the threat, community, provides a natural basis for organizing. The evidence on prior organization is mixed. In some cases, preexisting community groups and/or local political leaders provided a readily available organizational vehicle for opposition; in other cases, the toxics protest event appears to have been truly an innovation in a previously unorganized population.[53]

Considering, finally, macrosocietal conditions that are said to facilitate movement emergence, such as economic prosperity and a favorable political opportunity structure, one would have to conclude that conditions did not favor sustained and successful mobilization. In the early 1980s, just as grass-roots mobilization was taking off, the nation's economy was in recession. Recession was followed by deficit-fueled "pathological prosperity."[54] The economy seemed to thrive, but that prosperity, real or not, did not trickle down to the people who are active in the toxic movement, to working people with modest incomes, to African Americans, Latinos, Native Americans. *They* continued to experience a falling standard of living and the ever-present threat of unemployment—hardly economic circumstances that favor activism.

Turning to political opportunity structure: Yes, the existence of public participation provisions did provide an important focus and vehicle for organizing. In all other ways, however, the political opportunity structure did not seem particularly favorable. The overall tenor of the times was markedly conservative, hostile toward leftish notions. Even if the Reagan administration suffered through some embarrassing scandals toward the end of its tenure, one would be hard-pressed to argue that this was a time of regime crisis. Certainly, it was not a time when vulnerability made the state responsive to protest.

Yet, even though these larger societal conditions were anything but favorable, groups formed with increasing frequency; the movement thrived. That this movement experienced such vigorous growth even under adverse conditions underscores, I think, what can happen when media-produced issue-icons connect with felt experience and trigger feelings that one's most dearly cherished interests are at stake.

Movement Evolution

Movement emergence—people's willingness and ability to switch out of postmodern mode, to move from passively held beliefs and perceptions to active involvement—was the problematic and therefore conceptually interesting moment in the social movement phase of the hazardous waste story. After that transition, subsequent developments were, if not predictable, at least unsurprising from the point of view of what is known about social movement dynamics.

The development of more or less permanent infrastructural organizations is a routine feature of movement maturation. If a social movement is to survive past its first moment of spontaneous mobilization, it must be able to secure enough resources to keep going; "it must be able to generate support among authorities, sympathy among bystanders and, most important, an ongoing sense of legitimacy and efficacy among movement cadre and members" (McAdam et al., 1988:722). All these tasks are difficult, especially because they must be accomplished under conditions of scarcity and, most likely, overt hostility. That is why movements follow a predictable trajectory from more spontaneous actions toward the formalization of an infrastructure of social movement organizations (SMOs).

> Although collective action is expected to develop within micro-mobilization contexts, rarely are movements able to rely on them for their survival. . . . In most cases these micro-mobilization contexts may be little more than infor-

mal friendship networks, ad hoc committees, or loosely structured coalitions of activists. . . . For the movement to survive, pioneering activists must be able to create a more enduring organizational structure. (ibid.:714)

In contrast to infrastructure building, radicalization is neither a routine nor an inevitable feature of movement evolution. Movements follow a variety of different trajectories. Many, perhaps most, become more moderate over time.

How or why did the core infrastructure organizations in the toxics movement evolve toward the position I have called "environmental populism"? One can approach the question in two complementary ways. One can explore a functionalist explanation; one can, equally fruitfully, focus on *process*.

The functionalist approach directs our attention to the way a movement's ideological development is shaped by tactical requirements. According to McAdam et al., SMOs must constantly deal with the twin challenges of surviving in the larger, external environment of other organizations, while "continuing to mobilize the resources . . . they need to survive . . . [through] a continuous process of micro mobilization" (1988:716). An SMO's ideological development is, in effect, a function of tactical choices for coping with these tasks: "Goals and tactics are the *principal tools* an SMO uses to shape its external environment while simultaneously attending to the ongoing demands of micro mobilization" (ibid.; emphasis added).

Pursuing this type of analysis, one would point out that a radical populist rhetoric is well suited for reaching out to a rank and file that is confronting industry and government, becoming increasingly disillusioned, frustrated, angry, coming to disbelieve official depictions of what state and capital are all about. One would also suggest that issue expansion was desirable because it allowed the movement and its SMOs to expand their potential base of micromobilization settings.

Conversely, one would argue that there would have been little benefit in moving in the other direction, becoming more moderate. Moderating their rhetoric would have decreased core organizations' ability to forge solidarity with fledgling local militants. Besides, reform environmentalism is an organizational niche that is already well filled. Did the United States need one more organization engaged in traditional tactics of lobbying, of participating in the give-and-take of the official policy process? Carving out a new niche that combined direct, grass-roots action with an original and innovative environmental/populist ideology was the more promising, more functional choice.

The functionalist approach provides a useful, insightful perspective. It helps us think about the motivational context, the situational/tactical considerations that make it rational for a movement to develop in one direction and not another. This approach, however, does not tell us *how* that development occurs. For that, one has to switch to a process approach that draws attention to the specific experiences and interactions that lead key actors to articulate new understandings of their world and of their actions in that world.

Taking this approach to analyzing the development of radical evironmental populism, one would focus on the grass-roots activists who rose to positions of leadership and consider the experiences that would open them to new notions about their world, notions that would help them make sense of society as they were now beginning to see it. One would ask if the movement attracted people who had had previous activist experience and already had a well-developed radical analysis and, if so, seek to describe how their ideas influenced the others.

Anyone with the slightest interest in these matters knows of Lois Marie Gibbs, the leader of the Love Canal Homeowners' Association. Before the events at Love Canal, Lois Gibbs was apolitical, disinterested in the public sphere, a typical, rather diffident housewife. Today, Lois Gibbs is a strong, confident, public figure, a full-time activist and organizer, a leader.[55] Others, too, traveled this path. Having led purely private, apolitical lives, they became local activists and then went on to become leaders in the movement. In the following excerpts from interviews, three of them describe their initial experiences.

Sue Greer got involved when she found out that hazardous wastes were going into the landfill in her rural community, Wheeler, Indiana. She now heads PAHLS, one of the strongest, best organized statewide coalitions in the nation.

> We were just plain folks. I was a local lady. I lived there [Wheeler] all my life. My children went to school in the same school I went to.

> And what happened is: I was working at the local feed and grain store. And some guy came in, asked us if we were concerned about toxic waste being buried in a hole at the end of town. . . . everybody just looked at him because we really didn't know what he was saying anyhow.

> And the smell became outrageous, so I called my congressman. And I had never done anything like that before. I had to find out who the guy was, and then I called him and of course they go, they'll "have to check it out and get back to me." In the meantime, I thought, I gotta get an attorney.

> Well, I started getting all this information . . . all this outrageous information and I was just, like, horrified at it. . . . the information came into me slowly because it was a very technical, difficult subject for me. I mean, I grew up in farmland, USA, and I didn't know any of those terms, the words, the laws. . . .

> I got a letter from my congressman, the dear fellow, who said that they were only gonna dump about 500,000 gallons of this leaded tank bottom sludge from the refineries and it was all going to be over within about two weeks and it was really nothing to worry about. . . . So, we really thought that was a really ignorant thing for him to say. Now, I wasn't a college graduate. I didn't have any kind of Ph.D., but I surely was smart enough to figure out that that didn't sound like it was all going to be over. It might have been over for him, but I think that it was just beginning for me. I thought, "wow! If I take 500,000 gallons of something really ugly and nasty and I pour it in a hole, that's going to be a whole lot of nasty stuff and my water's under that hole."

> January 26, 1983, we had a special legislative hearing just for us—can you believe it?—in Indianapolis. The first time I was ever there in my life, Hal. I was 37 years old. I had never been there. I had never been to Indianapolis.

And I went there. I was terrified out of my skin. Well, it was three hours from my house. I don't think I ever went three hours away from my house. . . . when I was a little girl, we took like two vacations in my life, but I never went . . . that was way off my tether. The thing that I have to stress is that it was hard enough for me to understand all the words, all the new things that I had to deal with, but I had to deal with another problem, that of stepping off of this little circle that I was in and going out into the outer world. And that was real difficult for me. I had to struggle with that. And I think that if it wasn't for my motherhood-and-apple-pie gut feeling that if I didn't do something no one else would. . . . When we got down to the hearing Waste Management [the nation's largest waste conglomerate] people were there, because Waste Management owns our landfill. And they're no small peanuts and that was another shock for me to have to deal with because I thought it was Joe Blow and the dump and I found out that Joe Blow was much, much bigger and to have to deal with a humongous company like Waste Management was another shock to me. . . . I was traumatized by this whole thing. The fact that I had to step out of a whole secure world, I was very secure, and I had never done the things that I do today. . . . I had to do a lot of growing myself. . . . See now, nine years later, I can understand a lot of what I'm reading. In the beginning it was really Hell.

Q: Do you lobby?

A: No, I really don't. Number one, it turns me off because legislators, to me, are dishonest. . . . There are very few that I even trust. I think that if our forefathers came back and saw what was happening in this country they'd be shocked and appalled because it's not what their intent was. They have a bunch of people that are bred into corruption, they drink and carouse around, waste our money, they're greedy, they lie, they cheat, they have conflicts of interest. I mean, they're involved in multinational corporations, they cater to them and we are the losers for all those people.

Q: You were talking about your attitude towards legislators. I assume that developed since you got involved in this work.

A: Oh, it did. I didn't even know who they were before. Now I'm glad I know them for what they are.

We should be able to trust out government and the people that are running it to do the right thing for us. And they're not; they're doing the right thing for the wrong people.

Kaye Kiker became involved when an acquaintance gave her information about the corporation, Chemical Waste Management, Inc., that was planning to site a hazardous waste incinerator in her county. Kiker helped organize ACE, Alabamians for a Clean Environment. She is now the president of the National Toxics Campaign, one of the best known and most prominent leaders of the movement.

Q: Were you politically involved before you got involved with hazardous waste?

A: Oh, no. Nothing.

Q: What were you up to?

A: I worked . . . I was restoring my home. I was involved in social clubs around here, the Historical Society, the Order of the Eastern Star of Worthy Matrons, Garden Club, Homemaker Club. . . . Church: choir member and assistant Sunday school teacher. . . . I sewed. Made curtains . . . That was my life.

Q: I see. But nothing that had to do with politics?

A: No, not at all.

Q: And how about, were you reading newspapers and pretty aware of what was going on politically, or . . .

A: Well, like, I was more or less not interested. It just seemed like a lot of bad news. I didn't pay a lot of attention to the news. The newspapers—I really didn't read 'em. I was just too busy. Just always figured that, figured it was the same news, it just happened to different people. . . . Of course, local things was, interested me, ah, environmental things certainly wasn't, that was something, Love Canal was a long ways from my house.

I've always voted. [But] I wasn't the kind of person that was, in depth, um, you know, did any research about how people voted in the Legislature, what the mayors really do, you know. I had no idea. I just really trusted my government.

Q: How 'bout writing letters to congressmen, or going to City Council meetings?

A: No, never, not before ACE, no way, nope.

I remember Love Canal. I remember seeing that. I was concerned about the water and the land there. I thought it was an awful thing to do to the earth. I believe that the earth is a divine creation and a living thing. I thought it was a horrible thing.

It's like this: when it happens to somebody else, you think, "oh, those poor people. Ain't that awful." When it happens to you, you think, "I've got to do something." That's the way I felt. All of a sudden, it was in *my* backyard. . . . Trying to protect your own backyard. That's probably really what got me started.

I didn't want to get involved. A lady came to me. Said there was a dump in our county and they planned to burn hazardous waste there and would I help her to gather information to protect the community and I said I didn't want to get involved. I was sure the EPA was protecting us. And Governor Wallace wouldn't let something like this happen to us. I did not have the time. She asked me just to read some information about the history of the company and I told her I would. I told her I'd pray about it. The only thing I have ever been involved with speaking in public is the church. Anyway, I read it and prayed about it and felt a burden. I felt like it was something I needed to know about. And how dare they put poisons in our ground?

I didn't know at the time that it's the largest dump in the nation. I didn't know they were putting cyanide up there and deadly waste. I didn't know. We'd been there five years. . . .

We started from there, just gathering information. The more we found out, the more shocked we were. Realizing that, basically that people in our county did not have any knowledge and they didn't have any input in the decision.

The information was overwhelming. I was horrified that that was going on in my little county. I felt very protective of it.

[First meeting with officials:] I just listened because I didn't understand what they were talking about, anyway. Very technical. The terminology . . . is difficult to understand. I didn't know anything about environmental regulations . . . I was totally confused.

[Meeting with a group from Mississippi that was also fighting Chemical Waste Management:] They shared some information about Chem Waste with us and told us to get in touch with CCHW. Awful meeting for me. I just tried to absorb so much. I came home with a migraine headache. Stressful to learn what was going on in my county. It was unbelievable that the government would ruin land like that. Put deadly poisons in the ground.

I thought surely they [government officials] don't understand what's going on here. Once they understand it, and we get some information to them, maybe they can stop it [laughs].

When we first started, we thought all we had to do was tell the governor [George Wallace] what was going on and he'd stop it. Of course, we didn't know his son-in-law owned the site at the time.

[An employee at the landfill tells them he has been working with cyanide.] I was really alarmed. So I asked the state agent about that. He said that they don't dump cyanide up there. I said, "Is it allowed?" and the agent said, "No." I knew I had caught him in a lie. . . . The other lie came from the EPA. I called EPA to ask them about the history of the company. The EPA said they were just wonderful, hadn't done anything wrong, had a wonderful record everywhere. So that alarmed me, too.

Talked to city council, county commissioners, sheriff's department, tax assessors. It wasn't until we became pretty effective that they started ostracizing us. We went to the industrial development board in our county and we asked for copies of all their past minutes and they refused to give it to me. So we gave them a letter of request based on the Freedom of Information Act. They gave us copies ten days later that had been completely altered. One of the industrial board members called us and told us that we weren't given true copies of the minutes and to get the attorney general to notify them. So we were given original minutes then. Big difference. Left out information on the local attorney who owned the land who was also the attorney for the county, was the mayor of the city. Had a land deal of half a million dollars that was overpaid to him with county money. . . . The attorney was a member of this industrial development board. He had said in the minutes that anybody opposed to progress in Sumter County should be ostracized. The more information we uncovered, the more surprised we were, thinking that "this can't be happening."

I'm glad I was a NIMBY 'cause I know something I wouldn't have known otherwise.

Lew Dunn became involved when unpleasant odors, illness in his family, and other indicators convinced him that the hazardous waste disposal site in his community, Casmalia, California, was contaminating the area. Dunn now organizes other communities for Greenpeace.

Q: Before [you got involved with local water issues and the dump], had you been politically involved?

A: No, no.

Q: Not even in voting in elections?

A: No, no. I probably haven't voted in four elections in my life.

Q: What was your expectation, your sense of government before this and before Casmalia?

A: My sense of government was: we as citizens don't have to worry about anything because it is all taken care of for us. Hey, I got that senator up there and, shit, he's on Mount Olympus. Zeus.

Q: He wouldn't be a senator unless . . .

A: Oh, yeah. Unless he was the greatest guy in the world. How did he get to be my representative? 'Cause, I thought, everybody's smart, people are a lot smarter than I am and they elected him in office. He wouldn't do anything wrong or hurt me or that he would not protect me.

Q: And what would you have thought if something went wrong? Like a toxic waste problem, for example?

A: I never thought of it.

Q: How about this: If somebody unfairly gets harmed, is the government there to set things right again?

A: Oh, sure. Yeah. That it was an equal system. In other words it was equal for everybody. That's what I believed.

Q: And did you think that corporations were favored by government over citizens or that citizens had equal standing?

A: I really never gave it a thought, quite frankly. I never gave it a thought. My main concern in my life: I had no thoughts about GM, government or nothing. My main concern was raising my children and making sure that they had food on the table, that they had clothes and that if they wanted to go skin diving that I made enough money for that or if my wife wanted a new car or whatever, we could afford to have a nice car and a good place to live. That was my main concern in life.

I thought, "God, this is the greatest nation in the world. We have human rights that we are protecting in this country, and the government will take care of us and won't let people screw us and kill us."

I was intimidated by government officials. I wouldn't challenge them. I was fearful of them, to be honest with you. I feared making a mistake, saying the wrong thing. I didn't think I knew enough to challenge them. I never thought I would be able to do that.

In 1977, I had gone out to the site to take a load of trash out there and they said, "No, it's not that kind of dump." It said "Casmalia Dump" at the time. I said, "A dump's a dump." And then I just forgot about it.

Q: Were you smelling the stuff already?

A: No. We probably were but we didn't know what we were smelling. We weren't aware, really, that there was a problem at the time.

When I started to investigate what has happening in Casmalia, and I would dig up document after document after document that was showing that government didn't know any more than I did on how to solve problems or how to handle problems.

In 1981, I formed the first toxic advisory commission for the Santa Maria Valley in conjunction with the state and, at that time, I was naive enough to not understand the process. And they came and they would pump us for information, but we weren't learning anything. So I vowed, from that time on, if I went into a meeting, any government meeting, I had to feel that I came out with more than they did. Otherwise, I lost.

Q: What happened that made you aware of the dump?

A: About 1983, I guess. Well, we kept thinking the water was going to get contaminated because I'm digging up all these documents; shit, the state is coming up with documents, an incredible amount of stuff. I'm going through files and getting documents out and copying them and then I go back a week later, or a reporter goes back to get those same documents and they are purged from the file.

Q: Why are you digging up these documents?

A: Because by this time we are thinking about the water. We are hearing reports that animals are dying in the creek [from kids hiking in the hills]. In 1981, they find dead cattle in the creek and this is highly suspicious. . . . we went out there [Shuman Creek] with some news teams. Sure enough. There's all kinds of bones in the creek. . . . no crawdads or frogs or anything in the creek. It leads right from the dump. So I notify the Air Force [after the dump, Shuman creek flows through Vandenberg AFB]. . . . So, the Air Force comes out, with majors and captains and colonels and everybody. They tell us that they did an autopsy on the dead steer. They said that they didn't find nothing. . . .

Q: What did you do when they told you that . . . ?

A: I believed them. Why wouldn't I believe the Air Force, okay? . . . I trusted them, at that point. We were still naive then.

I went to [the county supervisor]. I'm talking about all this stuff and his administrative assistant . . . she gets really friendly with me. . . . I started feeding him [the supervisor] information about the dump and what was happening and I would sit in his home and he would tell me that it was terrible and he would just keep giving *them* the information. . . . I couldn't figure out how the site operator was getting the information. I mean, I was giving it to him [the supervisor] and the next day, they [the dump owners] had it. So, I figured there has to be a connection here. So I went to the county records and started digging up who were the owners [of the dump]. I found that his administrative assistant was one of the owners of the site. . . . all this time, I had been feeding her information.

The county air pollution control officer . . . was coming into our town when we were sick and our families were dying . . . our noses were bleeding, our neighbors were dropping dead like flies, and he was telling us that there was absolutely nothing wrong with this site and he was the one that would make the determination whether there was anything wrong or not. He now works for [the site operator].

This is when the heavy depression comes in 'cause you go to the doctor and they can find nothing wrong. You go to the county and there is nothing wrong. You go to the state. They say, "You guys are full of shit; we would not allow that to happen." And all the time, you're sick. You're neighbors are sick. All of a sudden, people start dying. This is a period of time, a two-year period when we lost seven people in Casmalia. . . .

Here we are, 1986, our neighbors are dying like crazy. They have been telling us for six years now that we are full of shit, that they don't believe us. They tell us that we are not smelling anything, then they come out and smell it themselves after it hit the attorneys and lawyers and doctors in their affluent homes. Only after that. What kind of trust would you have in the government?

I was outraged. You get to the point where, dammit, you cannot accept this any more. When you see your best friend blow up and die, I mean he literally blew up, there was blood everywhere. He was 51 years old. [Carol Dunn interjects: "He exploded."] Carol was with him. He exploded. He just sat down on the bed and exploded.

Why don't I see a toxic waste dump in Beverly Hills or next to the governor's mansion? Why do you take it out to rural communities?

I [once] thought, " . . . the government will take care of us and won't let people screw us and kill us." And you're naive to believe that. You're just naive to believe that because that is not what's happening in this country today.

I believe we have a corrupt system. . . .

I don't care what town, or what community or what state, if you have a facility that is operating with any type of toxic material, you'll find a problem because the pattern is there that industry and government just simply don't give a shit about whether they are breaking the law or not. . . . they want the money that that factory generates. Bottom line is: How much dollars can this company produce for the government?

Q: How did you come to this analysis of how government works?

A: Keep in mind, Hal, that the documents speak for themselves. I didn't create them.

In their previous lives, each of these folks had led overwhelmingly private lives filled with private, immediate concerns. They did not bother themselves with "political" matters. If they thought about that world at all, they pretty much believed in a textbook image of government; they trusted that officials do their jobs honestly and well. None of them was eager to get involved. At most, one hears them speak of doing it reluctantly, out of a sense of duty, because someone had to. Then, disillusioned and angered by their experiences, each moved toward a radical critique of society, business, and government.

Many of the movement's leaders have had exactly these kinds of histories. Local hazardous waste causes were their first political experiences. From the first, however, the movement also attracted others, people who had had experience in other movements. Some of these people came into the movement, I believe, because they thought neighborhood-based toxics organizing had mass movement potential. Others got involved simply because they, too, found themselves in contaminated communities or in communities where facilities would be sited.

The movement's key SMOs all have both types of activists in them, though the proportions may vary. NTC leadership seems weighted toward New Left radicals who came to grass-roots environmentalism by way of a preexisting, more general commitment to community organizing. CCHW leadership, on the other hand, prides itself on being made up of people who came up through the ranks after leading their own local struggles. Still, NTC has Kaye Kiker on its national board and CCHW has its Will Collettes.

Certainly, the presence of seasoned organizers who already had an articulate radical critique of American society had a big impact on how those newly disillusioned, newly active grass-roots folks would come to interpret what had happened to them and what they must now do. In an interview, Lois Gibbs confirmed that in the early days, when the "movement was very young," she "learned a lot" from colleagues who had previously worked in legal services and in the welfare rights movement. Further, she articulated how such people influenced the ideological development of the movement as a whole:

Q: Was there also an influence from people who have other movement experience who came to join the grass-roots environmental movement, in terms of passing along that way of thinking?

A: Yeah, I think that played a role. . . . They are trainers of the leaders, trainers of trainers. And then it has that domino effect, so then I go train somebody else and they go train somebody else. So they have sort of a major overall role . . . they're not out there training everybody, but they trained this group of leadership people who then consequently train everybody. . . . it's their influence on the leadership people.

It must be remembered, too, that the interactions between leaders rising from the ranks and seasoned organizers with previous experience took place in a cultural context in which radical/populist notions are a legitimate, if usually repressed and unacknowledged, part of the political culture and are always somewhat "in the air." Recall, finally, the situational/tactical pressures described in the "functionalist" analysis, above. Everything favored the development of a new hybrid, of a traditional left mass politics applied to a new grievance.

Conceptually, the movement's ideological evolution does not seem remarkable. Whether we choose to look at it in process terms, functionalist terms, or some combination of those two approaches, all these processes are well understood. The movement's radicalization is interesting and important not because it teaches us something new about "social movements" in general and in the abstract, but because of what that movement *is*, its historical originality and the impacts it has already had and might yet have in years to come.

The movement has opened up new tactical and conceptual directions for environmental action. Certainly, the movement did not invent direct, local, grass-roots environmental action—one has only to remember the first mobilizations against nuclear power plants, or Gladwin's research on environmental conflict in the 1970s. But the toxics movement made direct environmental action a mass, demographically diverse phenomenon. Certainly, too, this is not the only tendency in the environmental movement that has a radical critique of society. Here, one thinks of deep ecology and of ecofeminism, to cite only the most important examples. But, again, this is the only radical environmentalism that is truly a mass phenomenon.

Further, as I will discuss in chapter 8, the movement has recently embraced other social movements and has begun to say that it intends to join them, form alliances with them. This is in sharp contrast to other mass-constituency environmentalisms that have kept other causes, at least officially, in their public self-presentations, at arm's length.

Reactions and Consequences

These innovations have already had significant impact on the conduct of environmental politics and policy. They may prove to have political impacts that range far beyond the confines of a single issue.

In the next two chapters, I show that, in the short run, the movement provoked two quite contradictory responses. As siting opposition threatened to bring facility siting to a virtual standstill, policy experts, industry spokespersons, and government officials struggled, unsuccessfully, to find ways to neutralize this unruly assertion of popular power (chapter 5). At the same time, lawmakers could not help but be acutely aware of the depth of public sentiment. They responded by strengthening the federal laws that regulate hazardous wastes (chapter 6).

McAdam et al. tell us that "movement outcomes [are the] neglected topic" of the social movements literature (1988:727). Two subsequent chapters explore that question in some detail. In chapter 7, I argue that the movement's two seemingly contradictory policy impacts—motivating legislators to write stronger law, while acting locally to, in effect, disable implementation—have begun to force a major shift in the logic of pollution control policy, from a logic of "disposal regulation" toward a logic of "source reduction." In chapter 8, I describe how the movement has begun to reach beyond its "home" issue to embrace other social causes; I also address the ways in which it radicalizes movement participants.

Part II

Reactions

Chapter 5

Could Opposition Be Neutralized?
Discourses and Policies of Disempowerment

Industry and government officials, both, found grass-roots toxics activism, especially siting opposition, deeply disturbing. For industry, siting opposition threatened to create a disposal capacity shortage that would drive up their costs. For officials, local organizing threatened their administrative control over policy implementation. It would be far better, they agreed, if they found a way to secure communities' consent or, at least, their acquiescence.

Their concern gave impetus to an extensive, sometimes rather desperate-sounding, discourse: Could they find ways to contain and neutralize this troublesome exercise of direct popular power? Policy scientists dissected the causes of local opposition. They proposed a variety of solutions, ranging from direct preemption of local authority over land use to various compensation schemes to "expanded participation" models. But when the states, pressured by industry and the federal government to steel themselves against political pressures from below and *do something* to get the siting process moving, tried various combinations of these proposed solutions, the results were disappointing.

The Problem, as Regulators and Industry Saw It

Federal hazardous waste laws would, over time, have the effect of increasing demand for environmentally sound treatment and disposal sites. RCRA required that the hundreds of millions of tons of materials that had been dealt with, until then, in the easiest, cheapest, most haphazard ways, now had to be sent to proper, licensed sites. And things could only get worse. Without a systematic waste reduction policy, future economic growth would certainly mean even more wastes. As scientific data improved, more industrial waste streams would likely be added to the regulated list. Superfund would increase demand still more. Sites would have to be found for all that improperly disposed waste, the broken drums, the contaminated soil, all the contaminated by-products of the cleanup process.

At the same time, those laws would also tend to decrease the supply of facilities. RCRA was passed, in part, because existing disposal facilities were profoundly inadequate. RCRA's permitting process was *meant* to close down all those older, unsafe sites. When Congress passed the Hazardous and Solid Waste Amendments, the 1984 reauthorization of RCRA,[1] it put teeth into RCRA's permitting provisions. The new law dictated stringent standards that had to be met if a facility were to be granted a permit and set fixed dates by which facilities either had to apply for a permit or close their gates. In a few years, a significant fraction of existing disposal and treatment capacity would disappear.

Decreasing supply and increasing demand, together, meant that soon, if not immediately, industry would face disposal capacity shortages.[2] As discussed in chapter 4, regulators, industry officials, and spokespersons from the more established environmental organizations all agreed, albeit for somewhat different reasons, that insufficient capacity would be a serious problem. Some emphasized the economic consequences: generators were *already* having to go to great lengths to locate suitable disposal sites. The price of disposal was already much higher than it had been. If new facilities were not built soon, the impacts on economic activity would certainly get more serious. Others emphasized adverse environmental consequences: some generators would balk at the steep prices and the time and energy necessary to locate appropriate facilities. They would be tempted to cut costs and cut corners. "Midnight dumping" would increase.

It was essential, then, that new facilities be built. But making that happen would not be easy. In another time, perhaps, when people still believed wholeheartedly in technological progress, before society lost its innocence concerning the adverse side effects of modern industrial activity, building new capacity would have been straightforward and trouble free. Investors would have seen opportunity and submitted their proposals. Officials would have approved those proposals with little real scrutiny. Local citizens would have been acquiescent or, more likely, disinterested.

That, however, was no longer the situation by 1980, not after a decade of modern environmentalism and not after Love Canal and the subsequent spate of ominous stories about past waste disposal practices. The American people had experienced a profound "loss of faith . . . [in] government and private industry" (EPA, 1979:III).[3] People no longer unreflectedly believed that corporations act responsibly and regulators do their jobs well. They had come to assume the opposite, that neither private interests nor public officials could be trusted in matters vital to people's health or well-being.

If these attitudes did not change, news that someone was planning to site a hazardous waste landfill, a chemical treatment facility, or an incinerator nearby would almost certainly provoke opposition. Things would be different only if trust were somehow rebuilt and citizens came again to feel that private operators are responsible and government regulators effective.

But trust was not reestablished. Instead, as events unfolded, people's suspicions were repeatedly confirmed. Initial implementation of RCRA was a well-publicized failure. Those older, unsafe facilities continued to operate, now with the fig leaf of interim licenses. There were reports of midnight dumping and organized crime involvement in hazardous waste hauling and disposal. After 1980, trust in the capacity of regulators was further eroded by adverse publicity surrounding the Reagan administration's hostile and halfhearted administration of RCRA and Superfund.[4]

From the point of view of the average citizen, distrust of industry and of regulators, and hence opposition to their initiatives, seemed to make eminent good sense. From the point of view of industry, regulators, and most policy scientists, siting opposition was THE PROBLEM.[5]

What to Do?

Federal and state officials, consultants, the chemical industry, the waste disposal industry, lawyers, environmental organizations, policy think tanks, and social scientists all joined the debate about what to do. Some advocated *siting in out-of-the-way places*—industrial zones, rural areas—places away from neighborhoods that might oppose facilities. Others advocated siting in or near *communities that are most powerless*, least able or willing to organize effective opposition. Still others advocated direct disempowerment through state *preemption* of local control over land use. A fourth type of proposal suggested using *compensation* schemes that would secure local acceptance by altering host communities' cost-benefit calculations. Almost everyone agreed that, whatever else was tried, new forms of *enhanced participation* had to be developed.

Siting in Out-of-the-Way Places

Perhaps the simplest idea was to avoid opposition by siting in areas that are already heavily industrial, in underpopulated rural areas, or otherwise far removed from communities.[6] Siting in industrial parks would work because it would leave undisturbed existing land use patterns that people already accept. More generally, siting away from significant concentrations of population would work because—out of sight, out of mind—the new facility would be built away from people's gaze, away from the daily round of their activities. People might not even have to be informed. As EPA's consultants pointed out, "When the public is unaware of a siting attempt, they are unlikely to oppose it" (1979:18). Further, "when the site is in a heavy industrial area and not in the public view, a low-profile approach may be warranted. There is at least some evidence that opposition will not arise in these cases, so that there is no need to alert the public and thereby create a potential for opposition" (ibid.:24).

The approach seemed promising. Of the twenty-one siting cases Centaur Associates examined, "the four sites . . . which faced little or no opposition were all located in industrial areas" (ibid.:24). Cerrell Associates, a consulting firm hired by the state of California to study siting opposition, reported that "all in all, the most successful siting attempts of Waste-to-Energy facilities have been concentrated in industrial areas, preferably heavy industrial areas with little or no surrounding commercial or residential vicinities" (Cerrell Associates and Powell, 1984:22).

Still, the approach had its problems. Some facilities could certainly be built far away, but even remote sites often have *some* people living near them. If so, the strategy becomes one of "low-profile" or "silent" siting. Morrell and Magorian point out the troubling ethical and practical implications:

This approach carries a risk of backfiring and creating intense opposition [when it is discovered that] participation has been circumvented. . . . More importantly, however, the "low-profile" . . . [or] "[s]ilent siting" poses a special threat to the rights of residents near a proposed site. By choosing not to inform them of the future existence of a hazardous waste facility which may adversely impact them, decisionmakers deny these residents an opportunity to protest the imposition of these ill effects. (1982:126-127)

Selecting Communities on the Basis of Demographic Profiles

If one could not altogether avoid siting near human populations, one could think to enhance the likelihood of success by trying to site near those least likely, or least able, to resist. The most benign version of this approach is to try to select locales where people are politically conservative, probusiness, and still have a high degree of trust in experts and in technology.[7] Far more questionable is the strategy of selecting communities where people are socially and materially disadvantaged, powerless, and/or ill informed, communities that lack the resources, skills, confidence, and sense of entitlement necessary to organize resistance successfully.

The burden of living with industrial waste has always been unevenly distributed, falling more heavily on poor than on well-to-do, more heavily on black and brown than on white.[8] Although deplorable, at least this used to be, in a sense, an "unconscious" practice. It was the inevitable, though not necessarily intended, consequence of normal business logic, of both generators and disposal firms choosing the least expensive option in a regulation-free business environment. Industrially zoned areas, if close to residential areas, tend to be closest to poor neighborhoods, and, as noted earlier, almost all industrial wastes used to be simply dumped on the grounds of the plants where they were generated. Off-site facility location, too, used to be driven primarily by economic considerations, and that calculus tended to locate them in rural communities or in the poorest, most marginal urban environments.[9]

Now, however, faced with the prospect of burgeoning local resistance, some participants in the siting discourse advocated a *conscious* policy of identifying those communities least able to resist. Perhaps the most overt example of this can be found in a study done for the California Waste Management Board. In it, the consultants suggest:

> A demographic picture of the types of communities and the types of people that are most likely and least likely to oppose a Waste-to-Energy project would be invaluable to an effective siting program. A great deal of time, resources, and planning could be saved and political problems avoided, if people who are resentful and people who are amenable to Waste-to-Energy projects could be identified *before* selecting a site. (Cerrell Associates and Powell, 1984:17)

> Constructing a demographic profile . . . assist[s] in selecting a site that offers the least potential of generating public opposition. (ibid.:29-30)

The consultants found that the people "least resistant to major facilities" live in a rural community or small community of fewer than 25,000 people in the South or Midwest; are "old-timer" residents who have lived in their community for 20-plus years; are low income; are ranchers, farmers, or work in areas that are business related, technology related, or nature exploitative; have a high school education, or less; are above middle age; are politically Republican, conservative, and have a free market orientation; are not concerned about the environment; are not politically active or involved in voluntary associations; and are Catholic.[10]

There can be little doubt that, long before the Cerrell report made this approach explicit, practical-minded waste industry executives, in their efforts to select "site[s] that offers the least potential of generating public opposition," instinctively gravitated toward communities with a profile of "least resistant" people. Critics pointed out, however, that when this approach works, it tends to produce undesirable social justice and environmental outcomes. Choosing sites on the basis of weakest expected resistance produces "a not-so-subtle tendency to put these facilities in places where people ('the poor') will not object to them, rather than in places where they will do the least environmental damage" (Morrell and Magorian, 1982:116). In any case, students of public opposition warned that hazardous waste facilities were not like other types of LULUs; it would not be so easy to find pliant populations. The literature cites numerous instances of militant siting opposition in poor, working-class, and rural communities.[11] The EPA's consultants found that "public opposition involves a wide range of people . . . grandmothers and U.S. Congressmen, factory workers and university scientists, those who never graduated from high school and those with doctorates in ecology and physical sciences" (1979:III). Cerrell Associates themselves admitted in their report that

> the overriding conclusion of survey research on public attitudes toward these facilities is that opposition to the local siting of such a facility cuts across all subgroups. Regardless of socioeconomic status or residence, specific cases can be found in which even the subgroup least likely to form an opposition movement became intimately involved in the opposition struggle. (Cerrell Associates and Powell, 1984:18)

Preemption

Siting strategy could not be based solely on avoiding people. Nor could one be confident of finding communities that would not protest or could not do so effectively. Facilities had to be sited in or near communities that did not want them and could be expected to mount significant opposition to them.

Siting proponents were aware that citizen groups wield considerable political power at the local level. Local government bodies tend to be responsive to constituents' concerns and demands. In the United States, decisions about land use have traditionally been a local matter. Siting proponents worried that local governments would listen to outpourings of fear and anger, then use their authority over local land use to block siting attempts. "The most potent weapon available to facility opponents is the local police power, particularly the power to establish land use policies. Local government officials, reflecting citizen sentiments, may pass laws to exclude proposed facilities from their jurisdiction" (Bacow and Milkey, 1987:161). The answer seemed straightforward enough: state "preemption" of local control promised to neutralize local resistance by taking authority from the political institutions most responsive to it.

Preemption enjoyed a lot of early support. It was favored by "hazardous waste management facility operators, generators, trade association representatives and Federal and state officials" (GAO, 1978a:11). The EPA put a preemption provision into its Model State Hazardous Waste Management Act, and most of the early state siting statutes had preemption provisions.[12]

Critics soon argued, however, that preemption would not work, that its capacity to neutralize local opposition was a "myth."[13] Such a transparent attempt to bypass local sentiment would, they said, only make people more upset and more intransigent. Besides, local opponents would, in any case, continue to have other means, ranging from litigation to civil disobedience, at their disposal. They argued that, at most, preemption should be seen as only part of the solution and should be combined with other measures, such as compensation and expanded participation. They also insisted that preemption should take the more moderate form of an "override" provision that allows substantial local participation but permits the state ultimately to override if local "participation" turns out to be no more than a tactic to defeat siting through endless delays.[14]

Compensation

Others proposed that compensating host communities in some material fashion would improve their willingness to accept otherwise undesirable facilities. The problem, as they saw it, is that people are rational economic actors who will refuse to bear costs that exceed potential benefits. Like other LULUs, hazardous waste facilities offer important societal benefits, but those benefits are broadly diffused over a large region, while the costs are concentrated within the communities where those facilities are actually located.[15] If opposition is the result of people's rational weighing of costs and benefits, the remedy is obvious:

> The only practical response to this structural "tilt" in favor of local opposition . . . is to change local motivation to oppose. Compensation does this by reducing the costs each neighbor expects to suffer should the facility be built. (O'Hare et al., 1983:70)

> In theory, if the benefits obtained by the community from the project are increased so that they offset the residual social costs, the community should no longer have any incentive to oppose. Indeed, if the benefits to the community are large enough, it might actually desire the facility. (Bacow and Milkey, 1987:164)

> Compensation could take the form of direct money payments to local government

or to individual citizens. If monetary forms of compensation were unacceptable because that would be asking people to put a price on their own or their children's health, the owners of the proposed facility could offer nonmonetary forms of compensation. They could offer to mitigate effects most dreaded by neighbors, such as excess road traffic, noise, odor, local air and water pollution. They could offer property value guarantees or other innovative forms of insurance. They could offer valuable community services, improve roads, buy the community a new fire engine, provide free garbage disposal. If the land to be used for facility construction had previously been used by the community for other activities, such as outdoor recreation, the owner could provide other land where such activities could be pursued.[16]

Even if compensation would have worked as predicted, it would have been difficult to justify the seemingly inevitable social justice consequences. Although compensation need not only or even primarily take the form of direct money payments, every element of a compensation package can be readily represented in money terms. Economic logic dictates that facilities would be sited in communities that would accept them for the smallest total compensation package. Michael O'Hare, the most systematic advocate of this approach, says explicitly that compensation should be considered akin to an "auction," where the facility goes to the community that "bids" the least for it (1977:438). The equity problem arises because it seems inevitable that facilities would then all be located among the poor.

O'Hare addresses this issue head-on and defends the probable results. He agrees that it is a "plausible prediction that when noxious facilities are auctioned among towns of varying wealth, the poor towns will bid less and will acquire the facilities" (ibid.:453), and asks, "Why should rich people be allowed to buy their way out of their 'fair share' of regional disamenity?" (ibid.). He answers that redressing socioeconomic inequalities is a valid policy issue, but this is not the place to deal with it: "In general, we favor schemes to distribute income from rich to poor, but imposing a refinery on the residents of an occasional wealthy community is a clumsy way to do it, as are most redistributional schemes grafted onto policies whose fundamental purpose has nothing to do with income redistribution" (ibid.:453-454). Besides, who are we to say that the poor shouldn't have the right to improve their financial situation by accepting a disproportional share of society's toxic wastes? "From the point of view of the poor community . . . a government that prevents them from selling something at what they find an attractive price will not seem to have their interests at heart" (ibid.:454). And finally, O'Hare argues, "There seems little to recommend giving poor people an amenity (freedom from the refinery) they are anxious to sell, and only vengeance to recommend taking from the rich a good that is relatively worthless to others" (ibid.).

O'Hare's argument here takes on the cadence and spirit of the early English political economists, who, with great equanimity, argued that it was bad social policy to feed the poor.[17] A century ago, the poor had to be kept hungry so that they would be industrious and would be willing to serve in war; today, it seems, the poor are useful because they will sell, at a most favorable price, their willingness to bear more than their fair share of industry's toxic by-products.

O'Hare, champion of the inexorable logic of market rationality, was not fazed by the distributional implications of his compensation scheme, but, in a world where charges of "environmental racism," have been raised, a pure compensation scheme—baldly auctioning toxic facilities to communities most desperate for any additional source of revenue—was not really a viable political option.

In any case, studies soon showed that compensation, by itself, did not improve the rate of siting success in those states that had incorporated compensation schemes in their siting statutes.[18] Apparently, economic rationality was not a wholly adequate way to model people's attitudes and conduct in this situation. The subjective magnitude of the risks was so great that no amount of compensation could match it; or, contrary to purely economic theories of human action, people were not willing to monetize every aspect of their lives and were even outraged when asked to do so.

> It would appear that, in general, people assess the risks associated with living near a hazardous waste treatment facility as being so great that virtually no reasonable amount of compensation, by itself, can have much impact. Clearly, economic considerations do not seem to play the kind of role we would expect on the basis of the theory of compensation. (Portney, 1988:60)

> When I have discussed these ideas [financial incentives to encourage acceptance of the facility] with citizen groups they are affronted that the company would think so simplistically. While they are not totally unwilling to accept the possibility of such financial remunerations, they have no intention whatsoever of sacrificing the environmental quality of their community at the feet of mammon. (Windsor, 1983:7)

> It is only after safety issues have been resolved that the issue of compensation for the host community can be brought up. Compensation that is offered too early in the process is viewed by the public as a distasteful attempt at bribery. (Piasecki and Davis, 1987:180)

> Indeed, offers of compensation have occasionally increased local opposition; opponents of a proposed facility have attacked compensation as an immoral bribe. (Bacow and Milkey, 1987:165)

Compensation did find a place in siting policy, but always in combination with other methods, especially as part of expanded participation processes, respectful communications, and good faith negotiation.[19]

Expanded Participation

Public participation in policy implementation is a relatively recent innovation; by the mid-1970s, however, it had come to be seen as a normal and necessary feature of American political culture and was being routinely written into regulatory statutes.[20] Extensive public participation provisions were written into RCRA; each subsequent hazardous waste statute added more rights to have access to information, to participate, to litigate. Thus, even if people's perceptions were belittled as irrationally phobic by some, or too (economically) rational by others, they could not be excluded or ignored. Statutorally, the public *had* to be brought into the process. Trying to excluding people who already have a trust problem would, also, likely backfire.

> From a political standpoint, unilateral action, which forecloses the opportunity for public participation . . . is unlikely to accomplish anything constructive. If the public finds participation in the decision-making process foreclosed, it would simply pursue other avenues, most likely litigation. (McGuire, 1986:468)

> There is a strong sense that no facility can now be sited without the approval process establishing legitimacy. (Keystone Center, 1980:13)

On the other hand, industry officials and regulators were aware that simply following currently accepted participation formats would not work. First, it was often too little, too late. "Public participation" often consisted of no more than a hearing, held only after planning was far advanced. Policy analysts dubbed this format "decide-announce-defend" (Susskind, 1985:162) and pointed out that, to citizens, the usual public hearing seemed little more than an empty ritual that allowed them only the semblance of participation.

> [People get] the impression that the project is already underway, and that the citizen's groups are to serve only a legitimizing function. . . . if citizen's groups feel that their participation in the project is only to serve as a "rubber stamp," these same groups could become leaders of an effective opposition movement. (Cerrell Associates and Powell, 1984:32, 34)

> A fatal flaw in most governmental public participation is that it is grafted onto a planning procedure that is essentially complete without public input. Citizens quickly sense that public hearings lack real provisionalism or tentativeness. They often feel that the important decisions have already been made, and that while minor modifications may be possible to placate opponents, the real functions of the hearing are to fulfill a legal mandate and to legitimize the *fait accompli*. Not surprisingly, citizen opponents meet what seems to be the charade of consultation with a charade of their own, aiming their remarks not at the planners but at the media and the coming court battle. (Sandman, 1987:334-335)[21]

Second, given the intensity of public perception of the risks, such "participation" procedures would not only fail to achieve consent, they could also become settings where the opposition seizes the floor and uses it as a forum for dramatically presenting its views.

> The procedures for citizen involvement have been neither well thought out nor carefully applied. . . . the standard mechanism for involving the public—

the public hearing—routinely becomes a crowded, highly emotional exercise in mob psychology. (EPA, 1980:4)

Emotional bias and soapbox oratory often become the order of the day. (McGuire, 1986:468)

[Hearings] have [been] turned into political rallies. . . . It was how many people can you get into an auditorium to boo the speakers you don't like and cheer for the ones you support. (Sandman, 1987:335)

The hearing lends itself to "grandstanding," political posturing and use as a platform to build organizational strength and play to media. (Robbins, 1982:513)

EPA's study of public opposition showed that such concerns were not idle. Informational meetings and public hearings were frequently being transformed by angry citizens into sites for agitation and organizing.

How to steer between the Scylla of failing by giving people not enough voice and the Charybdis of failing by giving them too much? How could policymakers avoid disastrous extremes and find a course that could secure citizens' consent and make facility siting possible?

It seemed to many that the only way to restore trust and secure public acceptance was through fundamental rethinking and redesign of the process of citizen participation.[22] As Sandman put it, "[Not only is] genuine public participation . . . the moral right of the citizenry, . . . [a]s a practical matter, . . . public participation that is not mere window-dressing is probably a prerequisite to any community's decision to forgo its veto and accept a facility" (1987:335). What would need to be done differently? Examining the various prescriptions, "genuine participation" seemed to include (1) a change in *attitude* toward the community, (2) changes in participation *procedures*, and (3) willingness to make real *concessions*.

Before all else, attitudes had to change. Participation had to be authentically valued, not treated as empty ritual to be gotten through with as little trouble as possible. If one wished to achieve a community's consent, one had to approach it with respect, not patronizingly, not with the view that its concerns are irrational, not with some pejorative and dismissive notion of having to deal with "NIMBYs." Authentic dialogue meant a willingness to negotiate, to make the community a real partner in facility planning.

These attitudes had to be embodied in process changes that would involve citizens or their representatives fully at every stage of project development. People had to be brought into the process at the very beginning. They had to be given complete and honest information. Organizational forms had to be developed to facilitate the effective representation of community concerns.

Ultimately, expanded participation would be meaningful only if there was a willingness to make significant concessions. That could include mitigation of specific anticipated impacts, compensation for impacts that could not be mitigated, even a willingness to institutionalize a certain degree of continuing community control of how the facility would operate.

Unlike the other siting strategies, "expanded participation" appeared to treat opposition as legitimate in its own right. Looked at more closely, however, even this most democratic means of dealing with the situation ended up an uneasy, contradictory unity of concessions to and containment of direct popular power. Some of the techniques that were recommended in the literature on expanded participation were clearly co-optative in intent. For example, Robbins describes what a properly thought-out community advisory group can do: "If appropriately designed . . . carefully structured and well led . . . [a] task force . . . can be very useful in facility siting. It can . . . gain the support of community leadership . . . turn opponents into participants" (1982:514-515).[23] Robbins goes on to say that good results depend on picking advisory group participants with great care: "Selection of a task force is important. . . . members should be able to lead their constituent groups; but should not be true 'representatives' needing to return to the group for support" (ibid.:515). That way, the community will feel that the advisory group is legitimate, that the community leaders and opinion makers who serve on it can and will represent local concerns, but those "representatives" will not truly represent, in the sense of being appointed by and accountable to, the community or an organized constituency.

Ultimately, "participation" still meant different things to different actors. To movement participants, it meant the self-organized participation of direct democratic control over their lives. To regulators and business interests, it meant normalized, channeled, *proceduralized* forms of local input. Expanded participation was promoted with the rhetoric of citizens' rights in a democratic society; in the last instance, however, expanded participation intended *success*, not just better mutual *understanding*.[24] Community concerns had to be dealt with sincerely, but the ultimate point was still, and above all else, to make those sitings happen. That is why expanded participation procedures are often backed by the trump card of state override if participation does not secure the desired result. As the Keystone Center spokespersons stated: "If conflicts between local interests and site proponents cannot be resolved satisfactorily, however, the siting process should have provision for a state body to make a final decision. The 'threat' of such action should encourage mutual efforts to work out differences" (Craig and Lash, 1984:108).

The Policy: What Was Actually Done?

State government was the site where all this tactical discourse was translated into actual policy initiatives. The federal government had abdicated any direct role in siting. All the EPA could do was urge the states to recognize the need to make progress in this matter and move forward in spite of local opposition.[25]

By 1987, thirty-five states had adopted new siting statutes. These laws are complex, and there is considerable variation among them. Readers interested in specific details of these laws may consult a number of excellent surveys and summaries.[26]

Generally, the statutes avoid the most extreme options put forth in the debate on siting strategy. Undoubtedly, the disposal industry still finds notions of siting in out-of-the-way places or siting near poorer, more powerless communities attractive, but state siting statutes do not officially sanction such measures. Twenty-four of the thirty-seven statutes do contain preemption provisions, but mostly in the weaker form of state override authority. As Tarlock notes, "Most states have avoided up or down preemption choices and have sought preemption that accords the maximum possible local voice consistent with the objective of preemption—avoidance of local parochialism" (1984:440). Similarly, several states have adopted some type of compensation or other economic incentive scheme, but such measures are combined with negotiation and mediation strategies and are a far cry from compensation in the form of auctions soliciting competing low bids from poor communities. Several states have adopted versions of "expanded participation." This has taken various forms: Some states have put local representatives on state siting boards. Some have created local review boards and require facility developers to negotiate with them. Some give communities technical assistance grants so the communities can hire their own experts when they participate in the planning/siting process.

Finding themselves at the point where demands from all directions converged— pressured from above by the EPA and by industry to make siting easier, from below by communities that did not want to be made to accept other people's wastes—the states had been cautious and had reached for middle ground. Yes, they passed siting statutes, but they also made concessions to grass-roots sentiment by refusing to preempt local control fully and by writing some degree of host community input into their siting statutes. In addition, some states responded to citizen concerns by issuing at least temporary moratoriums on new siting; several states attempted to ban importation of out-of-state wastes.[27]

The Results: No Real Success

Given the distance, just described, between strong policy prescriptions and rather modest actual policy initiatives, it should not be surprising that public opposition was not neutralized. National surveys by two state environmental agencies and the EPA, as well as the annual surveys done by McCoy and Associates, an industry consulting firm, all tell the same story:[28] many new projects were proposed, but hardly more than a handful of new facilities were actually sited during the 1980s.

McCoy and Associates, perhaps the best single source of information on siting, reported that from 1983 to 1986, "the track record [was] truly dismal" (1986:4-1). Not a single facility was sited. Things seemed to improve after 1986, but, in fact, the apparent progress was illusory. A few new facilities came on line; a few existing facilities expanded; some cement kilns received approval to burn hazardous wastes, thereby increasing aggregate incineration capacity. Nevertheless, "the amount of new waste management capacity that actually became available was relatively small" (McCoy and Associates, 1990:4-2). Although no one seems to know the exact number of successful sitings that have taken place,[29] it is quite certainly far short of the 50 to 125 large facilities that EPA once claimed would be needed to avert a capacity crisis.[30]

At the same time, there were several waves of facility closures as older facilities failed to meet the more stringent HSWA standards. As a result, the McCoy surveys show, with the exception of high energy wastes that could be burned for fuel, total national disposal capacity *decreased* every year.

As the decade drew to a close, almost every attempt to site a new facility provoked the formation of a new "chapter" in the grass-roots toxics movement. Regulators and policy scientists were admitting that the siting opposition problem was not close to being solved. A tone of resignation, a certain fatalism of lowered expectations, had crept into the commentary. Consider some recent article titles:

"Public Relations and Participation: A Trail of Frustration with a Chance for Improvement" (Windsor, 1987)

"Siting Hazardous Waste Facilities in New Jersey: Keeping the Debate Open" (Dodd, 1986)

"Getting to Maybe: Some Communications Aspects of Siting Hazardous Waste Facilities" " (Sandman, 1987)

The last of these articles ends on this note:

Now is not the time to ask *any* New Jersey community to accept a hazardous waste facility. From "no" to "yes" is far too great a jump. We should ask the community only to consider its options, to explore the possibility of a compromise. Our goal should be moderate, fair, and achievable: getting to maybe. (ibid.:343)

Chapter 6

Hazardous Waste Regulation Progresses against the Conservative Tide

The political system had two very different reactions to icon and social movement. On the one hand, the officials who were charged with administering the hazardous waste statutes wanted to insulate themselves from the effects of the movement. As we saw in chapter 5, they applied themselves to the search for strategies that would neutralize what they saw as disruptive intrusions from below. At the same time, however, lawmakers felt compelled to respond to citizens' profound fear of toxic waste, their desire to be protected from it.

To get a sense of the strength of this latter reaction, one has to recall the larger political climate of the early 1980s. "Deregulation" was one of the battle cries of newly triumphant conservatives. The environmental and health and safety laws that were already in place would be under heavy assault. New initiatives, opportunities to strengthen various statutes when they came up for reauthorization, for example, would be avoided. Hazardous waste proved to be the only exception. Attempts to deregulate RCRA and Superfund caused a scandal. Congress then defied the deregulatory ethos of the times and significantly strengthened both laws.

Deregulation

Congress had responded to the rise of modern environmentalism, back around 1970, by enacting a broad array of new regulatory measures. In quick succession, it passed laws to protect workers (Occupational Safety and Health Administration, Mining Enforcement and Safety Administration), consumers (Consumer Product Safety Commission, National Highway Traffic Safety Administration), and the general environment (Clean Air Act, Clean Water Act, the Environmental Protection Agency).

The leading segments of the business community, the major trade associations, and spokespersons for the largest corporations certainly did not like the imposition of new regulations. But, given the conjunctural conditions, they saw that legislation could not be stopped. At best, they could lobby, persuade legislators that the laws they were about to pass ought be modest in scope. Following congressional passage, they would apply themselves to painstaking, detailed, systematic efforts to limit implementation and in this way lessen the regulatory burden.[1]

Circumstances were not right for a frontal counterattack and business leaders had assumed a defensive, damage-control posture. That did not mean, however, that they had resigned themselves to the new situation for all time. Recall Miliband's remark that although the state regulates business "precisely [for] the purpose of maintaining the right of property in general," owners and managers are seldom class conscious enough to appreciate that and are "seldom grateful for it." Conditions would change. Although a counterattack was not yet possible, one could prepare for the day when circumstances would become more favorable.

As early as the early 1970s, conservative economists began to develop the core claims that would justify future demands for regulatory rollbacks.[2] They argued that social regulation harmed economic performance. Regulation, they said, diverted investment, slowed the rate of technological innovation, hence slowed productivity growth. Too much regulation could not help but cause economic stagnation. Regulation hurt consumers because it raised prices. Regulation hurt workers because it caused job loss. The economists argued that national economic well-being could be assured only by giving the business world significant regulatory relief or, at least, by applying economic, cost-benefit criteria to regulatory decision making.[3]

Economic Downturn as Opportunity

The rationale for deregulation could be developed, but an actual campaign could not be launched unless conditions changed. The opportunity came soon enough. The oil price shock of 1973 and the recession of 1974-75 cast a pall on the latter half of the decade. The unemployment rate rose to 8.5 percent, a level not seen in a generation; at the same time, the rate of inflation rose to near 10 percent.[4]

The ailing economy did two things for proponents of deregulation. It made the case that something was indeed wrong with the economy. It provided an anxious, receptive audience for dire messages that past government policy, even if well intended, had been dangerously misconceived. In the late 1960s and early 1970s, Americans had expressed much less concern about the economy than about social issues and foreign policy. The percentage of people naming "the economy" as Most Important Problem rose precipitously in 1973 and eclipsed all other issues by late 1974. Polls that asked people about their expectations for their future personal well-being showed that people became more insecure and anticipated that things would get harder for them.[5]

The Discursive Campaign

Corporate spokespersons seized the opportunity thrown up by deteriorating economic conditions. They repeatedly told the nation that too much government intervention was, in great part, responsible for economic turmoil and that the economy would improve only if industry were given significant regulatory relief.[6] A cursory examination of various issues of *Vital Speeches of the Day* locates scores of keynote speeches organized around themes of big government, economic crisis, too much regulation:

"Abuse of Power by Regulatory Agencies," Z. D. Bonner, president, Gulf Oil Co., speech delivered at Town Hall of California, Los Angeles (1975, 41:194-197)

"Big Government or Freedom," William E. Simon, secretary of treasury, speech delivered at Kansas State University (1975, 41:386-389)

"Business and Government Controls," R. R. Lyon, Jr., senior vice president, Union Carbide, speech delivered to the West Virginia Manufacturers' Association (1975, 42:82-86)

"Human Liberties and Economic Freedom," Nelson A. Rockefeller, vice president of the United States, speech delivered to the Joint Economic Committee of Congress (1976, 42:386-387)

"Thoughts on Regulation," J. S. Smith, chairman, International Paper Co., speech delivered at the National Conference on Regulatory Reform, Washington, D.C. (1976, 42:663-666)

"Corporations and Public Opinion: Economic Freedom or Government Control," T. A. Murphy, chairman, General Motors, speech delivered to Associated Industries of New York (1976, 43:55-58)

"Business Policy and the Public Welfare: The Excesses of Government Regulation," Murray L. Weidenbaum, professor of economics, Washington University, speech delivered to the Executive Management Conference, New York (1977, 43:317-320)

"American Economy," Irving S. Shapiro, chairman, DuPont, speech delivered to the Southern Governors' Conference (1977, 43:738-741)

"Bureaucratic Babylon," W. L. Wearly, chairman, Ingersoll-Rand Corp., speech delivered to the Rotary Club, Syracuse, New York (1977, 44:44-49)

"Has Emotion Tipped the Scales on Consumer Safety?" J. W. Hanley, chairman, Monsanto, speech delivered to the Economic Club of Detroit (1977, 44:92-95)

"Regulation and the Public Interest," W. T. Yivisaker, chairman, Gould, Inc., speech delivered to the Public Relations Society of America, Washington, D.C. (1978, 44:604-607)

"Impact of Regulation on Business Development," A. Flamm, senior vice president, Union Carbide, speech delivered to the Commercial Development Association, New York (1978, 44:386-389)

"Impacts of Government Regulation," L. Loevinger, partner, Hogan and

Hartson, speech delivered as part of the ITT Lecture Series at New York University (1978, 45:130-145)

Articles attacking the economic costs of overregulation appeared in business-oriented magazines such as *Nation's Business, Forbes, Business Week,* and *Fortune,* in high-brow publications such as *Harper's* and *Saturday Review,* in the newsweeklies, *New Republic, Time, U.S. News & World Report, Newsweek,* and in mass outlets such as *Reader's Digest* and *USA Today.* A sampling of titles suggests the tenor of the articles:

"Getting Government out of the Marketplace," *Saturday Review,* July 12, 1975 (by Secretary of Treasury William Simon)

"Need to Reform Obsolete and Unnecessary Regulations," *Nation's Business,* June 1975 (by President Gerald Ford)

"Government's Hammerlock on Business," *Saturday Review,* July 10, 1976

"How to Halt Excessive Government Regulation," *Nation's Business,* March 1976

"Unlocking the Gilded Cage of Regulation," *Fortune,* February 1977

"What Businessmen Dislike Most about Government Regulation," *Nation's Business,* July 1977

"Federal Regulations: Catch-22 for Business," *U.S. News & World Report,* January 22, 1979

"High Cost of Regulation," *Newsweek,* March 20, 1978 (by Henry Ford 2d)

"Regulation Mess," *Newsweek,* June 12, 1978

"Stifling Costs of Regulation," *Business Week,* November 6, 1978

"Is Government Regulation Crippling Business?" *Saturday Review,* January 20, 1979

"Time to Control Runaway Regulation," *Reader's Digest,* June 1979 (by Murray Weidenbaum)

Antiregulatory agitation saturated public discourse. People heard the deregulatory message often, from reputedly knowledgeable experts and from political leaders. Polls show that they were increasingly willing to agree that it was true. There was increasing agreement that "government has gone too far in regulating business," and that "businessmen's complaints about excessive regulation" are justified.[7]

A more careful examination of polling data suggests that citizens' attitudes toward deregulation were, at best, ambiguous. When poll questions were phrased globally, when they asked respondents to produce an opinion about "too much regulation," in general, people faithfully repeated back basic deregulatory claims. Polls that asked, instead, about specific regulatory issues produced quite different results. Large majorities said that auto emission standards, standards for toxic chemical dumping, and the Clean Air and Clean Water Acts should be maintained or strengthened.[8]

Still, even if support for deregulation was superficial and equivocal, that did not much matter. There was more than enough evidence that anti-Big Government rhetoric had mass support. The advocates of deregulation could and did interpret that as wholehearted voter support for *their* interpretation of what "getting government off business's back" meant.

The Policy Campaign

A more direct policy campaign paralleled the public opinion campaign. Deregulation had advocates in every center of policy-making. In the White House, economists in the Office of Management and Budget (OMB) and the President's Council of Economic Advisers argued that the regulatory agencies had to be restrained if there was to be economic recovery. Congress's commerce and budget committees called for "regulatory reform." Corporations asked the federal courts to rule that regulatory agencies must consider economic impacts when developing standards.

The campaign for regulatory reform was beginning to bear fruit during the Carter presidency.[9] With the election of Ronald Reagan, the opponents of regulation prepared for final victory. Conservative think tanks and business groups, the American Enterprise Institute, the Heritage Foundation, the National Association of Manufacturers, the Chamber of Commerce, hurried to publish deregulatory wish lists. Murray Weidenbaum, on the Reagan transition team and soon to be appointed head of the Council of Economic Advisers, recommended a one-year moratorium on all standards activity, strict cost-benefit analysis, and enforcement relief. David Stockman, soon to be appointed to head the Office of Management and Budget, and Republican Representative Jack Kemp sent a memo to Mr. Reagan titled, "Avoiding an Economic Dunkirk," in which they warned that "regulatory evolution is just now reaching the stage at which it will sweep through the industrial economy with near gale force, pre-empting multi-billions in investment capital, driving up operating costs" (Stockman and Kemp, 1980).[10] The headline of an op-ed article in the *New York Times*, "OSHA; EPA the Heyday Is Over,"[11] expressed perfectly the heady sense that the time of defensive, containment tactics was over and the time for serious rollbacks had arrived.

The new administration made deregulation one of the three main planks in its economic program (the others being tax cuts and budget cuts). Executive Order 12291 required that all new and existing regulations be subjected to cost-benefit analysis. David Stockman was installed as head of the Office of Management and Budget, an agency that had accrued a great deal of power over the regulatory agencies in the executive branch. (It is relevant to recall that, while in Congress, David

Stockman had labeled RCRA a "monument to mindless excess"; Barke, 1988:152. Stockman was also a leading spokesman for the conservative faction most militantly opposed to passage of a Superfund law.)[12] Vice President George Bush was appointed to head the new Task Force on Regulatory Relief.

Conservative activists known to be hostile to the mission of the regulatory agencies were appointed to head the EPA, OSHA, and the Consumer Products Safety Commission. To head the EPA, for example, the administration appointed Anne M. Gorsuch, a former Colorado state legislator who had little environmental or administrative experience but had strong personal ties to James Watt and Joseph Coors, figures prominent in the conservative, antifederal, antiregulatory "Sagebrush Rebellion" movement. Rita M. Lavelle, appointed to head the agency's hazardous and solid waste program, came to the job after working in public relations for the Aerojet General Corporation, a company involved in dumping at Stringfellow, a soon-to-be-controversial hazardous waste landfill in southern California.[13]

Deregulating RCRA, Superfund

The breadth of the first Reagan administration's attack on the agencies that administer social regulations has been documented well and often.[14] Here, I wish only to recall some of what was done to the EPA, and to RCRA and Superfund, more specifically, in the name of "deregulation."

The administration's appointees acted as if they meant to bring every aspect of the EPA's work to a virtual standstill. The administration made deep cuts in the agency budget in fiscal years 1982 and 1983. Gorsuch declared that regulatory rule making must wait for adequate scientific research; at the same time, the EPA cut back its research effort. Enforcement was decentralized to the various states; at the same time, the EPA decreased the grants that supported state enforcement activities. Enforcement procedures were reorganized so frequently that, inevitably, confusion and inaction reigned. Many experienced professionals were laid off, and others fled the agency; those who remained suffered deep loss of morale.[15]

Although RCRA was still in its infancy, there was a sustained and systematic effort to undo what little had been accomplished in the previous four years. The agency deferred new requirements that facilities must have groundwater monitoring plans, must file quarterly groundwater quality reports, and must file annual reports. It deferred the requirement that facility operators carry adequate insurance to cover spills and other unplanned releases. It proposed that facilities operating with interim permits (almost all facilities, at that time) be allowed to expand by 50 percent without regulatory supervision. It stayed regulations governing incinerators and surface impoundments. The agency again delayed promulgating permanent standards for landfills and proposed to lift the ban on the disposal of liquid wastes in landfills.[16]

The minutiae of policy implementation can be numbing. It may help to step back and consider the combined impact of numerous, individually minor, actions on one type of facility. Take landfills, for example. Against all scientific evidence to the contrary, the agency continued to advocate landfilling as the disposal method of choice, "the lowest risk option currently available . . . a commonsense alternative to the indiscriminant practices of the past" (House, 1982a:96). At the same time, everything the administration proposed, all the suspensions and deferrals, would ensure that landfills would continue to be essentially unregulated. Landfills would continue to receive liquid wastes; they would not yet have to comply with interim requirements to monitor groundwater pollution, to have insurance, or to submit annual reports; they could expand up to 50 percent while operating under these meaningless, eviscerated interim permits; they would not have to come into compliance with more stringent, technical standards until the agency issued such standards, something the agency said would be delayed for several more years. No wonder a witness complained to Congress that "the director [of EPA] is taking us back to the days of the Love Canal" (ibid.:13).

Certainly, some of these moves were blocked. Incinerator regulations were reinstated, as were the financial responsibility rules. The groundwater monitoring and reporting requirements finally became effective when the deferral period ended. A crescendo of protest forced the agency to reverse its announced intent to allow liquids in landfills.[17] However, the administration cut RCRA's budget, by 23 percent the first year, cut RCRA grants to the states by 10 percent, and cut back drastically on RCRA enforcement activities.[18] Even if the administration was rebuffed on some of its specific deregulatory moves, the cuts, and the overall climate fostered within the Reagan EPA, were more than enough to ensure that RCRA implementation continued to flounder.[19]

In the case of Superfund, the deregulatory mission was carried out through a combination of delayed implementation and overt disregard for congressional intent. The law had instructed the agency to generate a national inventory of potential Superfund sites, do preliminary assessment, and publish a National Contingency Plan and a list of sites slated for priority action by June 1981. In 1982, the GAO reported to Congress that the EPA had accomplished none of this. The agency had lists of potential trouble sites, 10,300 sites on one data base, about 9,200 on another, but had not been able to develop anything like a true, reliable national site inventory. A list of 10,300 sites sounds impressive, but this was a very preliminary, tentative list. Over a third of sites on this list lacked even a preliminary assessment. The agency had no overall plan. Required to assemble a list of 400 priority sites by June 1981, the best the agency could do was produce an interim list of 115 sites, and the GAO charged that the ranking system used to develop this list was inconsistent, largely arbitrary, not a reliable instrument for identifying sites most deserving of immediate attention.[20]

The whole point of the Hazardous Substance Response Trust Fund, the working fund of $1.6 billion provided by Congress, was that it would allow EPA to clean up contaminated sites without having to wait until responsible parties could be found and be made to pay for the work. Apparently choosing to ignore the intent of the new law, the Reagan EPA resisted committing fund monies for cleanups. Instead, it proceeded in the traditional way—the way rejected by Congress as too slow and too uncertain to show results—first taking the time to try to identify responsible parties, then entering into protracted negotiations with them.[21] The GAO found that the administration was spending only a fraction of the funds appropriated by Congress:

> Lack of available funding for Superfund activities is not a cause of limited program accomplishments. In fact, Superfund obligations lag far behind the spending levels appropriated by the Congress. . . . As of March 31, 1982, the Treasury Department estimated that Superfund had a credit of about $303 million. . . . Actual EPA expenditures were $38.3 million, leaving a fund balance of about $265 million. (1982:7)[22]

The administration was "deregulating," in this case, by simply refusing to carry out the mission of Superfund in anything like a full or timely fashion. Not surprisingly, refusal to implement meant that the public was not getting the protection it so desperately wanted. Assessments of what was accomplished in 1981 and 1982 show amazingly poor results. Examining cleanup activity directly paid for by the EPA and the states, the GAO reported that, as of April 1982, sixteen months after Superfund was begun,

> few Superfund-financed remedial actions had been accomplished. . . . [there were some] remedial action activities at 40 of the 115 sites on the interim priority list. Cooperative agreements between EPA and the States had been signed for 11 of those sites . . . and 25 others were under negotiation. . . . [Money obligated at these sites included money for] 27 investigation/feasibility studies, seven engineering designs, and one construction project. (ibid.:6)

As stated, at this time, the EPA was emphasizing settlements to achieve private-party cleanups. But, according to its own figures, between 1980 and 1982, the EPA obtained only six settlements with private parties for final cleanups and another five for partial cleanups.[23]

Certainly, some cleanup activity was under way, but when compared with the estimated scope of the abandoned waste site problem, what was being done seemed hardly to scratch the surface, and it seems that not a single site had been totally and finally remediated.[24]

Deregulators Meet Their Match: "Sewergate" at the EPA

Deregulation, too, had its enemies.[25] Environmental organizations, consumer groups and labor unions, their congressional allies, professionals in the regulatory agencies were all dismayed at the prospect of what would happen to "their" agencies, "their" policies.

The substance of the administration's policy goals would have guaranteed active resistance, in any case; the Reagan appointees' style made that resistance even more intense. The appointees pursued their policies with abandon, audacity, undisguised contempt for those who disagreed with them. They routinely and even defiantly violated the norms of behavior that grease interactions in a social world in which people must continue to cooperate with present and future opponents. In interviews I conducted in Washington, they were described to me, variously, as incompetent, mediocre, indiscrete, stupid, inexperienced, occasionally malicious, and corrupt. The phrase I most often heard was that the Reagan appointees did not understand "how government works." As a consequence, interviewees unanimously agreed, the Reagan team at the EPA came to be despised by many of those who dealt with them.

Notably, the appointees alienated two groups with whom they should have, instead, sought to develop working relationships: members of Congress who oversee the regulatory agencies' work and those agencies' staffs of career professionals and scientists. The naturally strained relationship between agencies and oversight committees was made far more antagonistic than necessary by the Reagan appointees' undisguised contempt for the congressional oversight process. Members of Congress complained bitterly that the agency was ignoring their requests for information, and oversight hearings became increasingly confrontational.[26] Some agency professionals would have opposed deregulatory actions under any circumstances; others were alienated by the way these policies were being pursued. In the EPA, for example, staff were being laid off, and rumors of politically motivated "hit lists" circulated. The mood among the agency's employees ran a spectrum from demoralization to profound hatred. An oppositional staff culture developed. Dissident staff did what they could within the agency to frustrate policy initiatives with which they disagreed. They took resistance outside the agency, forging ongoing, semicovert alliances with congressional investigators, environmental groups, and the media.

Opposition to the administration's environmental policies began to coalesce as early as the Senate confirmation hearings for Anne Gorsuch. Senators and environmental organizations' spokespersons expressed sentiments ranging from "concern" to "dismay" at the prospect of seeing Gorsuch head the EPA.[27] Five months later, senators were raising questions about budget and personnel reductions, about rumors of poor morale at the agency, and about "signs of organizational deterioration."[28] Opposition intensified in 1982. National environmental organizations issued scathing reports.[29] More oversight hearings were staged.

Opponents did everything they could to block what Reagan's appointees were doing at EPA. There were similar efforts elsewhere as supporters of product safety, of worker safety and health, and of restraint in the economic exploitation of public lands tried hard to block what they felt was the undoing of a decade of regulatory progress. But none of it seemed to work. The deregulators pretty much had their way, in spite of oversight hearings, protests, repeated attempts to stir up bad publicity. The opposition had not yet found a way to really hurt the administration.

Toxic waste would prove the breakthrough issue. Love Canal was still fresh in the popular imagination. The public had been exposed to frightening TV images, witnessed the moving plight of contamination victims, been told repeatedly that hazardous waste was the nation's most serious environmental problem and that there were potential Love Canals in thousands of communities. Even if people agreed, vaguely, that "government regulation" had "gone too far," they emphatically wanted regulatory protection from the hazards of industrial waste. The media still found the issue compelling. Members of Congress had no trouble understanding that this was a special issue. They heard often from constituents who lived near Superfund sites, and "by the fall of 1983, approximately two-thirds of . . . congressional districts had at least one site on the Superfund National Priority List" (Harris et al., 1987:16).

Prelude to Scandal: The "Liquids in Landfills" Controversy

EPA's proposal to suspend a ban on the disposal of liquid wastes in landfills provoked a significant outbreak of controversy in 1982. Headlines that the EPA was seeking to ease rules for waste dumps and allow burial of barrels of liquid waste raised the specter of more Love Canals. Some members of Congress were quick to exploit the proposal's resonances. Congressman Levitas said, "They are simply talking about poisoning American people . . . an incredible act" (ibid.:39). Congressman Molinari asserted, "The trucks are rolling into 900 landfills all over America carrying a deadly legacy [for] our children and grandchildren" (ibid.), and Congressman Marks called the situation "an invitation to disaster" and said it "effectively removes all rules for the disposal of hazardous liquids" (ibid.).

As Harris et al. note, the proposal turned out to be "a political and public relations fiasco . . . [conveying the] devastating image . . . that 'hundreds of Love Canals' would be created. . . . Pressed to respond by the kind of media and political frenzy that rarely greets a federal regulatory decision, EPA was forced to acknowledge its miscalculation" (ibid, 39, 40). Once again, the iconic power of toxic waste was confirmed. The "liquids in landfills" controversy showed administration opponents that hazardous waste was the issue where they could break out of the restricted orbit of the Washington policy battlefield and generate a significant, nationwide political uproar.

Sewergate

In late 1982, two congressional subcommittees began to scrutinize the EPA's implementation of the Superfund. The committees believed, based on both public information and leaks by disgruntled agency employees, that the EPA was systematically defying the intent of the law. The agency was drawing from the fund as little as possible, declining to undertake cleanups of contaminated sites until agreement was reached with the polluting firms that were responsible. There were indications, furthermore, that in the handful of cases in which the EPA was reaching agreement with polluting firms, it was striking "sweetheart deals" highly favorable to the responsible firms. It was also rumored that Superfund head Rita Lavelle was manipulating cleanup timetables in order to influence the outcomes of important elections, blocking agreement on the Stringfellow site in California to hurt Jerry Brown's campaign for governor, hurrying to announce cleanup of a site in Seymour, Indiana, to help Senator Richard Lugar's reelection bid.[30]

The committees asked the EPA for all files concerning Stringfellow, Seymour, and several other Superfund sites. The EPA declined, stating that the files contained "enforcement sensitive" material that Congress could not be trusted to keep confidential and that premature disclosure would compromise negotiations with the responsible firms. In light of the allegations, it could be supposed that these particular files were special not because they contained enforcement-sensitive materials, but because they contained clear evidence of wrongdoing. Congressman Dingell stated that his subcommittee's staff "believes it highly likely that information relevant to the allegations of misconduct and unethical behavior by Federal employees is contained in the withheld documents" (House, 1985:460). Subpoenas were issued.

In response, the administration escalated and invoked the doctrine of executive privilege. Subsequently, official investigations suggested that some figures in the administration had been looking for an issue over which they could stage a showdown with Congress over the control of information; refusing congressional scrutiny of the EPA's Superfund activities was only the vehicle through which this confrontation was to be carried through.[31]

The decision to carry out this struggle over an issue as loaded, as potentially explosive, as hazardous waste proved a bad miscalculation. Since Watergate, claims of executive privilege have had the aroma of coverup. Many found it easy to assume that the administration was trying to hide serious wrongdoing, possibly even activities that were outright criminal. Furthermore, the executive privilege claim enlarged the terrain of the conflict from a struggle between a few subcommittees and the EPA into a struggle between Congress as a whole and the executive branch. On December 16, 1982, fifty-five Republican members of Congress, motivated as much by the issue of congressional rights as by particular concern about Superfund, joined the Democratic majority in voting a contempt citation for Gorsuch.

The Justice Department had instructed Anne Gorsuch to invoke the doctrine of executive privilege. Now, Congress was directing that the Justice Department prosecute her for acting as its agent. It refused to do so and, instead, appealed to the federal courts to void the contempt citation. The investigation of Superfund had suddenly blown into a full-fledged constitutional crisis.

The administration gambled that a favorable court order would both affirm executive privilege and bring congressional pressure on the EPA to a halt. But on February 3, 1983, District of Columbia District Judge John Lewis Smith, Jr., dismissed the Justice Department's suit and ordered the two sides to try to reach an accord. The decision left the situation open, without clear resolution.

Sensing the growing potential for political damage, administration officials moved to bring the situation under control. Rita Lavelle's "voluntary" resignation was announced the day after the court decision. Lavelle was a logical choice for scapegoating and ritual sacrifice. She headed the EPA program most directly under scrutiny and was already under investigation for perjury, conflict of interest, and harassment of a dissident at the agency. The administration also announced three separate internal investigations, by presidential counsel Fred Fielding, the FBI, and the Justice Department.

Sacrificial firing of lower-level personnel and promises of self-investigation are hallowed crisis-management tactics that often work to abort a fledgling scandal. Here, however, they did not do the job. Lavelle's firing went badly. It was alleged that her resignation note was written by superiors and issued to the press without her consent. Lavelle fought back, refused to fall on the sword, implicated others. Rather than relieving pressure on the administration, the tactic backfired and made things even worse.

Democratic members of Congress could feel they were gaining the upper hand. They intensified the pressure. New allegations—political manipulation of cleanup timetables, financial mismanagement, use of government resources for personal business—were added to old allegations of nonimplementation, conflict of interest, and sweetheart deals. News that paper shredders were at work at the agency implied that potentially incriminating subpoenaed documents were being shredded and led to fresh charges of coverup. Four more subcommittees announced hearings and investigations. More subpoenas were issued. All the charges, old and new, were amplified through daily repetition in press conferences.[32]

The administration tried again to abort the scandal. It offered to turn over the subpoenaed documents to Congress in a way that would save face by nominally preserving the administration's claim of executive privilege. They were ready to surrender, but why would the congressional Democrats let them off the hook when, for the first time in two years, they had a situation that could be leveraged into something big? The Democrats rejected the offer and pressed their advantage. The rhetoric escalated. The Democrats made new allegations. They charged that Attorney General William French Smith had wrongly interfered with the contempt of Congress prosecution and called for an investigation. Congressman Dingell claimed he had evidence of actual criminal misconduct.

Sewergate became an even bigger story than before.[33] The media dropped any pretense of neutrality. The *New York Times* editorialized about the "cynicism, mismanagement, decay" at the agency, called for Gorsuch's resignation, and quoted similar editorial demands from major newspapers around the nation.[34]

The administration's room to maneuver had shrunk precipitously. More firings of lower-level appointees were tried, but such moves could no longer bring things to a convincing or satisfying conclusion. The sheer duration of the uproar, the daily front-page coverage, had created an atmosphere of crisis, an aura of untrustworthiness. Inside the EPA, work had ground to a halt. The agency was in total uproar. Dissident staff gleefully discussed daily developments. Bets were placed on what day Gorsuch would resign.

The administration's facade of coherence began to come apart. Gorsuch (now Mrs. Burford) openly clashed with the Justice Department and White House staff. Even though she swore loyalty to Mr. Reagan the next day, the disorganization inside the administration could not be hidden. Republican members of Congress voiced concern about public confidence and suggested that Burford would have to go. A new poll showed that 54 percent of the public thought that the president cared more about firms that violate antipollution laws than about enforcing those laws.[35] White house aides "privately" (i.e., publicly, but anonymously) worried that the EPA was spoiling the "good news" presidency.

Sewergate had to be resolved before further damage was done. On March 9, 1983, Anne (Gorsuch) Burford resigned and the White House announced it would hand over all the documents wanted by Congress. The Democrats in Congress tried to maintain momentum by redirecting their attack toward appointees still in office, but the administration was now moving decisively. More resignations were arranged. The White House admitted that some allegations had been correct. The appointment of William Ruckelshaus, the first head of the EPA and someone with intact environmental credentials, to replace Gorsuch was both an admission that changes would have to be made and a tactic to get the EPA off the front page and, in the words of an anonymous adviser, "to kill the persistent notion that he [Mr. Reagan] sympathizes with big business on the environment" (Clines, 1983). These gestures left the opposition with nothing to do but savor the symbolic victory and voice cautious optimism that there would be real changes under Ruckelshaus. Media coverage faded as the story ran out of steam and the scandal episode drew to a close.

Scandal: Made Possible by the Power of the Icon

The appointees did commit questionable, unethical acts. Investigation revealed that they were cozy with regulated firms, gave personal assurances of nonenforcement, agreed to sweetheart deals that allowed polluting industries to avoid full payment of cleanup costs, and manipulated waste site cleanup timetables to affect Republican candidates favorably in congressional races. Several appointees failed to sequester themselves from agency action involving former employers. No matter how deplorable such behaviors were, however, they do not seem criminal enough, on their own merit, to be the inevitable raw material of scandal. Even after the crisis boiled over, after months of uncontrolled leaking, congressional scrutiny by six subcommittees, and intensive investigatory reporting, the *New York Times* could report that "no evidence clearly establishing that any crime has been committed has yet been made public" (Taylor, 1983). Nor were these behaviors markedly different from what was being done with the rest of the EPA's programs, with OSHA, with the Consumer Products Safety Commission, or with the Department of Interior.

The Reagan appointees' administrative style certainly contributed to their downfall, as well. Their behavior made enemies of both members of Congress and agency staff, who then gladly joined environmentalists in mounting an ever more effective opposition. But was their behavior so different from that of the appointees at other regulatory agencies? The tactless, unrelenting, uncompromising fashion with which they pursued the administration's policies seems to have been fairly typical of the overall style of the early Reagan years. That the EPA administrators alienated other important actors in Washington explains those actors' *motivation* to join the organized opposition, but it does not explain their *success*.[36]

Administration opponents tried to generate public outcry, tried to block the work of the deregulators on a number of fronts. In the case of worker safety and health, for example, labor unions and organized labor's congressional allies strove mightily to arouse public disapproval and block what OMB, the Bush task force, and new agency head Thorne Auchter were doing to OSHA.[37] Similar efforts were made over consumer products and over the policies at the Department of Interior, with little success.[38]

By trial and error, the opponents of deregulation found out, between 1980 and 1983, that they could take on the administration and win on one issue and one issue alone. It is no accident that administration opponents succeeded only when their agitation concerned the nation's two toxic waste statutes. The recency of the iconogenic process gave the threat of toxic waste special urgency and immediacy. With toxic waste, "deregulation" ceased to be an abstraction to be vaguely approved as contributing to economic improvement. Deregulation was, instead, experienced as the very real dread that one could find oneself in a position like that of those people at Love Canal. The opposition found, first with the "liquids in landfills" controversy and then with Sewergate, that they could orchestrate something of a replay, the passionate coverage, the outcry, of that earlier iconogenic moment, and that they could wield significant political power when they did so.

Scandal Confirmed and Reinforced Power of the Icon

Sewergate was a top political story for five months. Some political careers were ruined. Rita Lavelle went to jail. A popular president's administration suffered serious embarrassment, even if the landslide electoral victory enjoyed by Reagan a mere year and a half later suggests that his loss of face during Sewergate did not cause him permanent damage.

What impact did the scandal have? In terms of immediate policy payoffs, the EPA's overall performance improved somewhat. Ruckelshaus recruited assistants who were generally more competent, more experienced, and more professional than their predecessors. Staff no longer had to strategize against a hostile cadre of bosses in order to carry out even routine duties. Order and morale were restored and the situation inside the agency stabilized. These were real improvements compared with the dismal state to which the agency had sunk.

Specifically with respect to hazardous waste activities, the agency declared that Superfund and RCRA work would be a top priority.[39] The Superfund budget tripled in two years, from $210 million in fiscal 1983 to $640 million in fiscal 1985; Superfund staffing doubled.[40] According to Ruckelshaus, every aspect of the Superfund program—emergency response short-term cleanups, investigation, and engineering studies at long-term, Priority List sites—accelerated. By 1984, the EPA had 546 sites on the National Priority List and Ruckelshaus could proudly say that "some level of activity is under way at nearly every site" (1984b:288). One perspective, then, is that the scandal led to significant gains over what had been done under the Gorsuch/Lavelle regime. This was certainly the message broadcast by the agency's own public relations apparatus.[41]

Still, even if it was true that activity was under way at many Superfund sites, it was taking years to move from initial assessment and engineering studies to actual cleanup activity.[42] In 1985, the EPA reported that of the 110 sites on the original, 1981, National Priority List, cleanup activity had been completed at five sites, was partially completed at thirty-nine sites, and was under way at nine others.[43] Also in 1985, the GAO examined the status of work at the 538 sites then on the NPL.

> As of December 31, 1984: Thirty-six percent or 194 of the 538 priority sites had no cleanup action underway or planned; 44 percent or 236 sites are in the investigation and/or study phase; and 19 percent or 104 sites had Superfund-financed or responsible party cleanup action approved or underway. EPA considered cleanup activity complete at the remaining four priority list sites. We reviewed files in detail for the 58 sites approved for cleanup as of June 30, 1984. . . . We found that most — 47 sites — involved planned actions which would only partially or temporarily resolve the sites' problems. . . . *The 10 sites EPA considered cleaned up generally involved relatively uncomplicated remedies compared to problems EPA currently faces at most NPL sites.* (1985d:2, 6; emphasis added)[44]

GAO studies of the EPA's RCRA activities during the second half of the decade suggest a similar pattern: a far more serious effort than during the Gorsuch years, but persistent, serious problems in all phases of implementation.[45] One could argue equally plausibly, then, that the scandal accomplished precious little, if judged not by contrast to what had been done before but in relation to the scope of the environmental problems the regulations were meant to address.

I would like to argue, however, that the impact of Sewergate cannot be measured by tables of site activity or other indicators of implementation alone. The scandal process, the saturation media coverage, the opinion polls, and the resultant damage done to reputations and careers forcefully reminded Washington of the enduring power of the hazardous waste icon. Even if we agree that the scandal produced only limited improvements in the short run, it enhanced the power of those in Washington who were fighting to strengthen the hazardous waste laws when, a few years later, those laws came up for reauthorization.

Reauthorization of RCRA, 1984

Before the scandal, efforts to reauthorize RCRA were stalled, even in the face of incontrovertible evidence that its original design was seriously flawed and that the effort to implement it had been a dismal failure.[46] An unusually broad array of interested parties — scientists, environmentalists, congressional liberals, government researchers, state officials, even the hazardous waste treatment and disposal industry — had come forth to denounce the EPA's performance.[47] If things continued, they warned, there would be more contaminated communities, more citizens distraught about their children's health, more Love Canals.[48] To no avail. The House did pass a reauthorization bill in 1982 but, in the Senate, controversy regarding a rather minor issue, how to deal with "small generators," firms that generate less than 1,000 kg/month, was enough to stall the bill in committee.[49]

After Sewergate, however, the climate for RCRA reauthorization improved markedly.[50] Congressional liberals felt that conditions had become so favorable that they could now do much more than simply get the statute renewed. They proposed to strengthen every facet of the existing regulatory scheme. They ventured further still and proposed several true innovations in how regulatory laws had traditionally been written.

First, the language of their bills was unusually detailed: "In the breadth of their coverage, and their extraordinary detail, these revisions read in many places more like a package of new regulations issued by an executive agency than a piece of legislation" (Mugdan and Adler, 1985:216). Observers agreed that "the level of technical detail specified by Congress . . . is unprecedented in environmental legislation" (Harris et al., 1987:101), an "unmistakable deviation . . . from Congress's approach to other environmental legislation in recent years" (Mugdan and Adler, 1985:217).

Second, the bills contained self-enforcing deadlines that came to be known as "hammer provisions":

> Perhaps the most important innovation of the 1984 RCRA amendments is the incorporation of self-executing regulations. In the event that EPA fails to meet any of the numerous statutory deadlines imposed by the amendments, so called "hammer provisions" automatically become effective. Thus, if EPA does not act within the time frame specified by the statute, statute itself specifies the regulation that becomes effective as of the deadline date. (Ottinger, 1985:22-23)

These innovations would fundamentally alter the power relationship between Congress and the regulatory agency. The fact that they were proposed shows not only that congressional liberals were distrustful of and impatient with the EPA,[51] it shows their new confidence, their sense that, with Sewergate, the political situation had changed.

While scandal had strengthened the hand of the bill's proponents, it weakened the position of its opponents. In 1982, eleven of the House Energy and Commerce Committee's eighteen Republicans had opposed reporting even a quite weak reauthorization measure; in 1983, after Sewergate, the committee voted unanimously to report H.R.2867. The *Congressional Quarterly Weekly Report* commented, "The lack of negative votes so far this year may indicate that members are taking shelter from the political fallout of recent controversy on the Environmental Protection Agency's handling of hazardous wastes" (April 30, 1983:840). Even the handful of members of Congress still willing to oppose overtly some of the reauthorization bills' provisions felt it necessary to keep affirming that they did agree that RCRA was not protecting the public health and did need to be revised.[52]

For its part, the Reagan administration signaled that it was willing, in spite of some internal dissent, to go along with almost anything. The Office of Management and Budget objected to the "inflexible and unnecessary regulatory mandates," but "other officials, including certain White House staffers and EPA Administrator William D. Ruckelshaus took a more positive view . . . , seeing it as a bill President Reagan would be well advised to sign in an election year" (*Congressional Quarterly Weekly Report*, July 28, 1984:1817). Ruckelshaus said that he did not "like the deadlines," but that he considered them "inevitable because Congress does not trust EPA. The best aspect of the bill, Ruckelshaus said, is that it should, for several years, end the squabbling between the executive and legislative branches on how EPA should be controlling hazardous wastes" (*Environment Reporter*, 1984, 15:955). In fact, when things got bogged down in the Senate,

> White House staffers and Ruckelshaus made calls to senators July 24 urging compromises to smooth the way for passage of the bill. . . . lobbyists specu-lated that some White House officials concluded it was in Reagan's political interest to sign such a bill if it was going to pass anyway, and that the White House had been looking for a means of deflecting some of the environmental criticism that Democrats have been hurling at Reagan. "We are looking at a Rose Garden signing ceremony for a major piece of toxic substance control legislation three weeks before the election," a Senate Republican staffer said. (*Congressional Quarterly Weekly Report*, July 28, 1984:1817-1818)

Examination of the legislative history shows that the reauthorization process was neither simple nor straightforward,[53] but the bill signed by Mr. Reagan on No-vember 8, 1984, retained most of what the proregulatory forces had wanted. Again, the administration's enemies' ability to foment scandal confirmed the special status of the toxic waste issue. This is how we understand the fact that RCRA reauthori-zation became "the only major piece of environmental legislation to emerge from the four years of the 97th and 98th Congresses" (Mugden and Adler, 1985:215).

Reauthorization of Superfund, 1986

Superfund reauthorization followed much the same pattern. Superfund's sponsors aggressively promoted major changes. The size of the fund ought to be expanded, they argued, to $10 billion or more. In a move that echoed what Congress had done in the case of HSWA, members of Congress argued that the new law should dictate to the EPA detailed standards that site cleanups must meet and that the law should set timetables for the number of cleanup actions the EPA must undertake.[54] They tried to reinstate certain citizen rights provisions that had had to be abandoned to get Superfund passed in 1980, toxic victim compensation and creation of a "federal cause of action" that would make it easier for victims to pursue legal action against polluters directly.

Conservatives opposed much of that, of course, but the scandal made Superfund's opponents suddenly much more conciliatory, and their postscandal position was far different from their position before. Before Sewergate, the Reagan administration had planned to oppose reauthorization of the Superfund program in any form. The glacial pace of implementation under Ann (Gorsuch) Burford and Rita Lavelle was, in part, motivated by the desire to spend as little of the original $1.6 billion as possible, so that the administration could go to Congress and argue that the program needed no new appropriations. The scandal forced the administration to make an abrupt about-face. In his 1984 State of the Union message, President Reagan announced he would support reauthorization of Superfund.[55] In fact, the administration proposed to *expand* the fund to $5.3 billion.

The reauthorization process itself was, if anything, even more confusing and complex than the reauthorization of RCRA.[56] Proponents could not turn their advantage into total victory, and they failed to secure every provision that they wanted. Still, environmental organizations considered the final bill a "major victory" (*Congressional Quarterly Weekly Report*, December 14, 1985:2619). The fund had been increased to more than $9 billion. The new law contained "unusually specific" cleanup standards and timetables that "curtailed . . . EPA's discretion . . . in several significant respects" (Hedeman, Shorb, and McLean, 1987:193). Although the victim rights provision again had to be abandoned, the new law strengthened citizens' capacity to sue EPA, mandated local public participation in Superfund cleanups, and included a new title, the Emergency Planning and Community Right-to-Know Act of 1986, that significantly increased public access to industry toxic emissions data.[57]

Powerful voices in the Reagan administration, the Office of Management and Budget, and Treasury Secretary James Baker criticized the size of the fund and the tax scheme and counseled a presidential veto, but the risks were too great. The bill's supporters warned that they would put a very damaging spin on a presidential veto: "I would urge and hope the president would sign it. I don't think he wants to go down in history as being in favor of hazardous waste dumps" (Senator Stafford, in *Congressional Quarterly Weekly Report*, September 21, 1985:1894). Eighty-one Senators urged Mr. Reagan to sign the bill. Fifty-seven Senators wrote Republican Senate Majority Leader Dole asking him to keep the Senate in session to prevent a pocket veto.[58] Mr. Reagan could not ignore what Leslie Dash of the Audubon Society called "the plaintive cries of Republicans up for re-election" (*Congressional Quarterly Weekly Report*, October 18, 1986:2624). Senator Stafford has noted that "the president appreciated that there was strong national support for the bill, that people were scared and wanted something done. And I think his political advisors told him that it was the wise thing to do, both morally and politically" (ibid.). Although his signing statement made plain his distaste for the bill he had in hand,[59] Mr. Reagan signed the Superfund Amendments and Reauthorization Act on October 18, 1986.

Part III

Results

Chapter 7

Fifteen Years of Hazardous Waste Legislation: Summing Up the Policy Impacts

All attempts to make the hazardous waste movement go away, to neutralize or co-opt it, had failed. "Toxic waste" continued to evoke feelings of dread. Recognition of the issue's continued importance and the movement's undiminished power gave Congress the will to defy the tenor of the times and strengthen both RCRA and Superfund.

What do legislative accomplishments mean, however, when laws are chronically underimplemented? The two hazardous waste laws have never been anywhere near fully implemented. Their history appears once again to confirm everything that social scientists have said about regulatory failure.

I wish to argue here that the traditional policy outcome perspective is too narrow. Traditional measures of policy impact are important, certainly, but we must consider things more broadly. For one, regulations have promoted significant infrastructural development. More important still, legislation and grass-roots activism have, together, created a situation that is forcing a shift away from a dependence on *pollution control* toward the more desirable logic of *pollution prevention*.

Narrowly Defined: A Continuing Failure

If one wished to evaluate regulatory impacts narrowly, one would have to conclude that fifteen years of legislative effort have yielded very disappointing results. The histories of RCRA and Superfund implementation are histories of unrelieved failure, although the *causes* of that failure have changed over the years.

With both RCRA and Superfund, the problems inherent in starting any new program were made worse by indifferent or outright hostile presidential administrations. The Carter administration supported RCRA, but considered it a low-priority, expendable item. Proposed standards drafted by EPA staff were weakened when Carter's economic advisers argued that they would be inflationary. RCRA budgets were held extraordinarily low, lower even than the unrealistically low amounts appropriated by Congress.

Though things improved after Love Canal, the thaw lasted only a couple of years before the 1980 elections put RCRA in the hands of an administration that was thoroughly unsympathetic to its mission. Superfund, passed by Congress only in the last days of the Carter administration, was in the hands of unfriendly administrators from its first moments, with predictable results.

Following the political embarrassment of the Sewergate scandal, the Reagan administration proclaimed its eagerness to protect Americans from the threat of toxic waste. Budgets increased somewhat. Under friendlier administrators, the regulatory bureaucracies began to function once again. It seemed that things were moving forward again. Better yet, the reauthorizations contained implementation-forcing provisions with which Congress intended to limit the EPA's administrative discretion, to assure that implementation would continue to improve.

Then, even as these developments raised hopes that the regulatory effort would improve, larger conditions intervened and imposed new constraints on how far those improvements could really go. Worries that the federal deficit was growing out of control meant that restraining budgetary growth had become a top priority in Washington. Federal programs struggled just to keep from being cut; certainly, conditions did not favor significant program expansions. So, even if implementation was marginally better than before, the still immense gap between what needed to be done and what actually was being done could not be narrowed significantly.

Not surprisingly, recent evaluations of the quality of RCRA and Superfund implementation have continued to be highly negative. The GAO has issued a stream of sobering reports about the RCRA program after more than a decade of regulatory efforts:

Hazardous Waste: Uncertainties of Existing Data (1987c)

Hazardous Waste: EPA Has Made Limited Progress in Determining the Wastes to Be Regulated (1986b)

EPA's Inventory of Potential Hazardous Waste Sites Is Incomplete (1985a)

Hazardous Waste: Facility Inspections Are Not Thorough and Complete (1987b)

Hazardous Waste: Groundwater Conditions at Many Land Disposal Facilities Remain Uncertain (1988a)

Hazardous Waste: Many Enforcement Actions Do Not Meet EPA Standards (1988b)

Illegal Disposal of Hazardous Waste: Difficult to Detect or Deter (1985c)

Hazardous Waste: Environmental Safeguards Jeopardized When Facilities Cease Operation (1986a)

Hazardous Waste: Corrective Action Cleanups Will Take Years to Complete (1987a)

Recent evaluations of Superfund, conducted by Congress's Office of Technology Assessment (OTA), by the RAND Corporation, and by a coalition of environmental organizations and the Hazardous Waste Treatment Council, are equally bleak.[1] All agree that, after nearly a decade of Superfund activity and after billions of dollars spent, the EPA had succeeded in cleaning up only a small fraction of the sites that require attention.

Infrastructural Development

Given this history of grossly inadequate results, it is not easy to dismiss those who would conclude that having good laws on the books is essentially meaningless. Certainly, from the point of view of average citizens' immediate concerns with safety and well-being, regulatory laws are meaningful to the degree that they actually ensure that facilities do not pollute, that contaminated sites are cleaned up, and so on. Still, such a surface reading of regulatory outcomes misses real accomplishments.

Beyond the specifics of implementation, the very presence of a regulatory statute fosters the development of "issue infrastructure," that complex of knowledge, technology, and institutions that makes it possible for society to understand and cope with any issue. Recall, for the moment, the description in chapter 2 of the conditions that existed before RCRA. The United States had practically no infrastructure for proper control of hazardous industrial wastes. There were almost no data on waste generation, disposal, or the toxicity and health impacts of wastes. There was no real hazardous waste disposal or treatment industry; industrial wastes were (mis)handled by the same trucking firms that hauled household garbage. Mostly, wastes were simply thrown into the ground. The few environmentally adequate disposal facilities that did exist were underutilized. In the absence of regulation, there was no incentive to build new capacity or to do the basic research to develop better treatment technologies.

Much of that has now changed. The statutes mandate the production of information about all facets of the problem. Consequently, society's knowledge, although far from complete, is much, much better than it was fifteen years ago. There is better understanding of what chemicals are being dumped, where and in what quantities, their toxicity, and their impact on the environment and human health. The public information and public participation provisions written into the statutes ensure that much of this information is made available to the public.[2]

The past fifteen years have also seen the development of a real disposal and treatment industry. Before regulation, the "industry" consisted mostly of simple hauling companies and landfills. Today, the United States has a modern, multibillion-dollar disposal industry.[3] Although it is true that even the biggest and best firms in the industry have poor operating records—the EPA's enforcement efforts, though admittedly anemic, find many facilities are in chronic violation of RCRA standards[4]—the overall trend is positive. The worst of the old facilities are closing, going out of business. An active research program is developing new and promising treatment technologies.[5]

True, implementation has not been good enough really to protect public health and the environment, in the short run. And, undoubtedly, a stricter regulatory climate would have stimulated even more infrastructural development. Neither of these facts should cause us to deny that, even with inadequate implementation, there has been real improvement in society's overall ability to manage industrial wastes.

Beyond the Limited Logic of Regulation: Toward Pollution Prevention

Important as it is, the development of treatment and disposal infrastructure serves only to improve how toxics are dealt with once they have been created. I wish to argue, next, that there has been a policy impact that is potentially far more momentous. The complex interactions among public attitudes, regulation, and social movement—what I have called "issue history"—have begun to force a turn from the logic of pollution *control* toward the superior logic of pollution *prevention*.

Recall the RCRA legislative history described in chapter 2. Once it had been agreed that the nation's toxic waste streams had to be managed better, how would the government do that? Would it create a tail-end control system that would attempt to regulate wastes once they had been generated? Or would it force industry to make changes at the point of production, thereby significantly reducing the amount of wastes generated in the first place?

Given that states in capitalist societies tend to avoid strong interventions into production, it should not surprise us that, bowing to intense industry pressure, Congress shied away from controls on the generators. Once again, the solution would be better control, not prevention. The source reduction alternative seemed to have suffered a decisive defeat. Yet, by the end of the 1980s, both government and industry officials were praising source reduction as the most desirable, possibly the only workable, solution to the nation's waste problems.

Policy Moves toward Source Reduction

In policy circles, the shift was discernible by the mid-1980s. There was a flurry of conferences and official reports. In 1985, the National Research Council sponsored a conference on waste reduction and published *Reducing Hazardous Waste Generation: An Evaluation and a Call for Action*.[6] Congress's Office of Technology Assessment issued its report, *Serious Reduction of Hazardous Waste*, in September 1986; the EPA followed with its own report in October; and the OTA issued another, *From Pollution to Prevention*, the following June.[7]

In the Hazardous and Solid Waste Amendments of 1984, Congress declared that source reduction would be the goal of national policy; the act contains several provisions that prod industry to move in that direction.[8] By 1988, seven waste reduction bills were pending in the 100th Congress.[9] Not content with the pace of progress at the federal level, by 1986 about twenty states had enacted waste reduction laws more advanced than what had been achieved by Washington.[10]

The reasons policymakers would distance themselves from their earlier hard-line rejection, back in 1976, of the source reduction alternative are not hard to discern. The standard regulatory approach had proven to be a fiasco: unworkable, unpopular, a continuing political disaster.

It was increasingly clear that neither RCRA nor Superfund was working. Prospects for ever achieving full implementation appeared bleak. Worse, several careful studies had shown convincingly that even the best landfills eventually fail and will inevitably pollute groundwater, threatening public health and the environment.[11] Many had begun to conclude that RCRA's disposal regulation scheme would not work *even if* it were fully implemented.

The American people desperately wanted lawmakers to do something to protect them from toxic wastes. Elected representatives were besieged by constituents angry because they feared their local contaminated sites were not being dealt with adequately or because someone was trying to site new facilities in their communities. In these circumstances, it was not hard for lawmakers to listen to the scientists and the environmentalists and conclude that embracing waste reduction was both the right and the smart thing to do.

One should not exaggerate what had taken place. The 1976 victory against waste reduction had been eroded, not reversed. Private sector spokespersons were still adamant about government interference in production decisions. David Graham of Dow Chemical Co. said that "industry's No. 1 concern is that waste minimization be voluntary, not mandatory" (in *Environment Reporter*, 1988, 19:175). Another spokesperson stated, "Direct control by government over the production process in industry is unthinkable" (Editor, *Hazardous Waste and Hazardous Materials*, 1987). Public and quasi-public expert evaluations agreed:

> Policies that directly control industrial processes through regulations would be extremely complex from an administrative and practical point of view. The sheer number of and variations in industrial processes throughout the country make effective administration of a program that specifies required changes in industrial processes very difficult. Moreover, waste reduction involves changes in manufacturing processes, which have generally been outside the traditional purview of environmental regulations. (Committee on Institutional Considerations in Reducing the Generation of Hazardous Industrial Wastes, 1985:18)

> Mandatory waste reduction regulations . . . could harm international competitiveness for some industries and products because they are too inflexible, are inattentive to site-specific constraints, or ignore capital investment needs. . . . standards applied equally across U.S. industry could have grave consequences for troubled manufacturing sectors. (Office of Technology Assessment, 1986:23)

Thus, even as state and federal policy increasingly promoted waste reduction, it observed the structural/historical limits on state intervention and policymakers continued to disavow direct government controls on production. For example, although

HSWA required generators to certify that they were reducing generation "to the degree practicable," Congress put on record that this provision does not authorize anyone "to interfere with or intrude into the production process or production decisions of individual generators" (Harris et al., 1987:165).[12]

The Private Sector Embraces Waste Reduction

In 1976, the major players in the private sector acted as if the very idea of source reduction was anathema to them; at the least, it was strictly their own business and they wanted no public discussion of it. In contrast, by the mid-1980s, the leading elements of the private sector seemed to have developed a positive passion for waste reduction.

There were numerous conferences and seminars.[13] Some of the nation's largest corporations, such as DuPont, 3M, AT&T, Hewlett-Packard, and Union Carbide, to name just a few, sent senior representatives to these conference or published articles in the journal *Hazardous Waste and Hazardous Materials*, to describe proudly what their companies were doing to reduce waste generation.[14]

How real was all this? Industry spokespersons were, of course, of the opinion that "a great deal has already been done" (Editor, *Hazardous Waste and Hazardous Materials*, 1985a). We should approach such claims with caution and assume that they are, in part, strategic verbiage to keep regulators at bay.[15] To get a balanced picture, I reviewed surveys by the National Research Council, the Office of Technology Assessment, the New York public interest organization INFORM, and the EPA, and consulted the annual hazardous waste industry surveys carried out by McCoy and Associates.[16] Taken as a whole, these studies tell us that private sector movement toward source reduction is in its earliest stages. It is uneven and far more likely to be found in the larger firms. But the trend is undeniable and it is significant. Although waste reduction is not in any real sense required by law, many firms are moving toward waste reduction more or less "voluntarily," as it were.

What Caused This Shift Toward Waste Reduction?

Why have firms taken up waste reduction even though they have not been directly forced to do so? I begin with some official explanations.

The National Research Council cites five reasons:

> substantially higher costs for disposal, owing primarily to the strengthening of regulations;
>
> prospects of substantially greater liabilities for remedial (cleanup) actions, owing primarily to judicial and administrative interpretation of the liability provisions of the 1980 Superfund law;
>
> risks of third-party liability, even when a generator is not directly responsible for the improper disposal;
>
> potential for adverse public relations; and

public opposition to hazardous waste management facility siting (Committee on Institutional Considerations in Reducing the Generation of Hazardous Industrial Wastes, 1985).

The Office of Technology Assessment provides the following reasons:

regulatory programs have raised costs, compliance costs, the price of insurance, and the costs associated with the cleanup of polluted sites;

regulatory programs (especially Superfund) impose liabilities that increase the likelihood of civil and criminal prosecution, make it harder to secure adequate insurance, and potentially affect every industrial real estate transaction;

other private sector institutions, insurance companies and banks, may require waste reduction before providing liability coverage or loans; and

environmental organizations, public interest groups, and grass-roots organizations have made waste reduction a priority (1986, 1987).

The EPA cites six "strong incentives" that "already . . . promote waste minimization in the private sector":

state and federal regulations have increased costs;

difficulties in siting new facilities, owing to "intense public opposition," have begun to create capacity shortages;

more stringent permitting and corrective action requirements make getting RCRA permits difficult, expensive, and time-consuming; the resulting delays in providing new capacity have temporarily driven costs still higher;

financial liabilities stemming from both common law and the "imminent and substantial endangerment provisions of Sections 106 and 107" of Superfund have been felt;

shortages of liability insurance have been expensive ("In recent years, premiums have increased 50 to 300 percent . . . many companies have difficulty obtaining coverage [required by RCRA and HSWA] at any price"); and

public perceptions of company responsibilities have shifted; companies want to minimize wastes "out of sensitivity to public concern over toxic chemicals" (1986).

Such lists are not wrong, or even incomplete. What they lack, however, is an analysis of the dynamic linkages that synthesize the individual items into a single, coherent historical narrative. Missing that, such lists cannot isolate and identify the cause or causes that were ultimately responsible for the new interest in waste reduction.

In contrast, the issue history provided in earlier chapters highlights the connections among these factors or causes. That issue history suggests, I wish to argue, that industry's interest in waste reduction is, in the end, attributable largely to the complex and contradictory impacts of the community-based hazardous waste movement. At the risk of some repetition, let's review the relevant points.

In theory, the passage of RCRA should have begun to raise the price of disposal, driving subpar facilities out of business and requiring a higher, and more expensive, standard of care. However, the fact that hazardous waste had not yet become a real mass issue weakened RCRA's potential impact. Delays in standard setting, unsupervised interim licensing, and lack of a sustained inspection and enforcement effort meant that older practices persisted.

Hazardous waste finally became a well-formed issue between 1978 and 1980, when extensive media coverage crystallized it as a stable image-object in popular perception. For the first time, lawmakers and regulators felt it necessary to take RCRA seriously. In the immediate aftermath of issue creation, Congress also passed Superfund. Superfund created liabilities that were much more exacting than existing, common-law liability standards. EPA interpreted strict liability as meaning that " 'responsible parties' are liable under the statute *even if* they can prove their conduct was lawful, in good faith, nonnegligent and consistent with the applicable state of the art industry and government standards" (Chesler, Rodburg, and Smith, 1986:9-10; emphasis added). Joint and several liability meant that even if only a small fraction of the wastes at a particular site came from a particular corporation, that corporation could be held liable for cleanup costs for the whole site. Furthermore, the statute authorized punitive triple damages and penalties.

As of 1980, then, laws were in place that could raise the cost of proper disposal and impose severe penalties for improper disposal. Together, these laws could potentially provide economic incentives that would make it rational for generators to take a serious look at the waste reduction alternative. However, the Reagan administration's commitment to deregulation prevented the realization of this potential.

At this point, the hazardous waste movement intervened and pushed things to a new level. It did so by affecting things simultaneously at two different levels of the political process. At the centers of policy-making, the movement's presence created pressure for legislative progress. At the same time, the movement acted in a decentralized fashion, in a multitude of communities, to hinder facility siting.

At the Center of Policymaking. Members of Congress understood what the polls confirmed: toxic waste was the most feared of environmental problems.[17] They repeatedly heard from constituents who were either fighting to defeat a proposed siting or concerned that they had a local Love Canal.[18] Remember, by 1984, more than two-thirds of congressional districts had a site on Superfund's National Priority List. Perception of the issue's salience emboldened the administration's opponents in Washington. Successful scandal-mongering put them in an excellent position to influence Congress as it began to reconsider its approach to hazardous waste regulation. As a result, Congress found the will to defy the conservative, deregulatory political climate of the early 1980s and greatly strengthened both RCRA and Superfund when they came up for reauthorization.

HSWA's ban on landfilling, its stringent requirements for facility licensing, its instruction that noncomplying facilities close, and its imposition of postclosure requirements all had the effect of reducing capacity and raising costs. Its "hammer" provisions ensured that the EPA could not water down these impacts by delaying or refusing to implement Congress's intent. SARA expanded liability to cover any real estate transaction where Superfund response costs are subsequently incurred. Industrial sites are bought and sold all the time. This extension of liability created new incentive for private parties to monitor each others' hazardous waste management activities to ensure that they would not get stuck with someone else's Superfund cleanup liability.[19]

At the Local Level. Stronger regulation would have increased the cost of disposal in any case, but perhaps that increase would have been more modest if supply could have kept up with regulation-created demand, that is, if the disposal industry could have responded by rapidly building capacity (and, in the dreams of the generators, overcapacity). But this was not to be. Almost every proposal to site landfills, incinerators, and the like met with vehement opposition. Because, as I showed in chapter 5, efforts to neutralize local opposition have not worked, facility siting has been practically at a standstill. Thus, while regulations force demand, locally, people vetoed the supply, and the price for legitimate disposal rose even faster. Land disposal, for example, used to cost about $15 per ton. By 1986, the cost of land disposal had risen to $250 per ton and the cost of incinerating wastes to $500 to $1,500 per ton.[20]

My argument, then, is that the hazardous waste movement is responsible for the progress toward waste reduction, but not in the simple and direct sense that pollution prevention is now one of its major demands. Rather, the movement is responsible because it created something like a "scissor" effect: at the centers of formal political action, the movement caused regulations to be strengthened. Locally, the movement threw a wrench into the siting process, making it nearly impossible to build new disposal and treatment capacity. The *combined* impact, the movement's contradictory presence simultaneously at these two levels, is, I believe, the principal historical fact behind each of the individual items that are said to cause industry's "voluntary" move toward waste reduction.[21]

If so, I believe the negative evaluation with which I began this chapter needs some qualification. Narrowly conceived, in terms of policy outcomes alone, the issue history appears to have produced no more than the usual, disappointing results. However, looking at the issue history more broadly, as we have just done here, suggests, instead, quite a different conclusion. The core logic of regulation itself has been made problematic, its weaknesses starkly revealed, pressures for a different answer mounted. The hazardous waste movement did not invent the idea of source reduction; it was, however, the historical agent that created the conditions that finally forced that idea to the center of environmental policy. To the degree that we believe the standard logic of regulation to be a failure and a dead end, this is a significant accomplishment.[22]

Just Saying No: On The Rationality of Mass Refusal

Not long ago, there was an almost religious faith in the connection between science and technology, on the one hand, and progress and material improvement, on the other. Over the course of recent decades, that faith has eroded. People have acquired a profound distrust of technological developments. It is not only hazardous waste facilities—people now presume that many industrial activities threaten their health and their well-being.

Those who still wholeheartedly believe in progress through technological innovation are exasperated. To them, popular, lay perception of the risks is obviously exaggerated, irrational. Take, for example, the comments of Chauncey Starr, author of a seminal article on risk.[23] Keynoting a symposium on societal risk assessment, Starr told the audience that "the perceptions [of the 'great population of this country'] may be so far from reality that you and I know that they're absurd, but that's how they feel about it" (1980:4). In a recent article, Starr cautioned that "public fears can always be aroused by the concept of man's tampering with nature, creating global catastrophes. Such fears can easily be used to stop new scientific developments. . . . anticipatory arousal of public anxiety can inhibit the creation of new technology" (1985:98-99). Or take well-known political scientist Aaron Wildavsky. He accuses Americans of suffering from unfounded fears, mass paranoia:

> How extraordinary! The richest, longest-lived, best-protected, most resourceful civilization, with the highest degree of insight into its own technology, is on its way to becoming the most frightened. . . . It isn't much, really, in dispute—only the land we live on, the water we drink, the air we breathe, the food we eat, the energy that supports us. Chicken Little is alive and well in America. (1979:32)

Charges of irrationality are frequently leveled against hazardous waste activists. Recall the discussion of NIMBYism in chapter 4. People, it is said, wildly overestimate the risks of toxic chemical wastes. They cannot distinguish between the irresponsible disposal practices of the past and the safer, technologically advanced, tightly regulated disposal methods of today. Futhermore, people's unreasonable militancy produces irrational *results*. Economic performance suffers; both nature and public health are threatened when firms cannot find convenient, reasonably priced disposal services and, desperate, resort to illegal and unsafe alternatives.

There are, I believe, two plausible ways to answer such accusations. One could, first, agree that the charges are correct, but point out that this is only another manifestation of the contradiction between individual rationality and societal, collective irrationality, a contradiction that is characteristic of, hence endemic in, capitalist societies. Marx long ago traced this contradiction to the very core of the logic of capital. Individually, factory owners act in their rational self-interest when they increase the rate of exploitation, lengthen working hours, substitute machinery for human labor, cut wages, and so on; collectively, the sum of their individual actions produces crisis conditions and cyclical downturn. In today's world, every time corporations resist regulatory control, challenge regulations in the courts, complain that environmental protection hurts their capacity to operate efficiently (also, let's not deny it, every time workers and unions argue against environmental preservation because they have families and they want their jobs), the rationality of sacrificing the collective for the sake of the immediate and personal interest is taken as self-evident.

In the case of hazardous waste, we can grant that there may be an incompatibility or contradiction between personal rationality ("No one knows how dangerous this material really is; it is impossible for me to know that this operator is reliable; often they aren't; regulators can't guarantee my safety. Therefore my best bet is not to have the facility sited close to my home.") and social rationality (it has to go somewhere or the ecosystem will be polluted). But, if this contradiction is fundamental to the capitalist organization of society, why should communities slated to host hazardous waste facilities be blamed for reproducing it, for failing to live up to a standard of social responsibility seldom seen in this society? Accusations of selfishness ring hollow in a culture steeped in privileging individual interest and starving the public sphere in every phase of social life.

Furthermore, accusations that putting self before society is selfish and irresponsible and assertions that if one enjoys the benefits of modern production one must also be willing to bear its costs are suspect unless the accusers are willing to argue for siting a *disproportionate share* of hazardous waste disposal facilities in communities that consume a disproportionate share of the nation's goods and services—in Westchester County, New York; Lake Forest, Illinois; Bel Air, California; Shaker Heights, Ohio; Leona Helmsley's and Charles Keating's backyards; and the rest of the American Haute Bourgeois Archipelago.

Such thoughts lead directly to considerations of equity and fairness, to analysis of the differential distribution of the toxic burdens of modern production. This is quite a fruitful avenue to explore in a world where polluting facilities are disproportionately sited near disadvantaged communities, where the advocates of compensation schemes to gain community acceptance freely admit that such methods inevitably mean more siting in poorer communities because they are cheaper to bribe, where consultants advise states on the demographic profiles of communities least likely to resist facility siting. This intersection of environmentalism with the politics of class and race is becoming ever more important, but further discussion here would divert me from the main points I wish to make.

In summary, the first way to answer the charge that local toxics actions result in undesirable societal outcomes is to agree, then ask, Why blame people for following the standard decision rules of capitalist society? But this answer is not entirely satisfying.

A second, more positive, response challenges the movement's critics on *every* one of their charges: people in the movement are neither irrational in their perceptions nor narrowly selfish in their motivations; the movement is forcing changes that are the epitome of long-term rationality.

Are People Too Afraid?

The first charge, that opposition is based on irrationally exaggerated fear and inability to distinguish between safe and unsafe facilities, is not difficult to refute. The claim that people are overly fearful assumes, first, that the risks are known; if the true magnitude of the risk is not known, how can fears be said to be exaggerated? In fact, research has yet to establish the degree of health hazard associated with the presence of improperly disposed hazardous waste.[24] The claim that the risks of *proper* disposal or treatment are known rests on the assumption that permitted facilities operate as advertised, a claim that is not credible in light of the EPA's enforcement record. Given the current state of knowledge and the current state of regulatory enforcement, there is no way to validate claims that the risks are known to be minor or acceptable, and therefore no basis for making blanket claims that people are "too" afraid of hazardous waste.

The claim that people are unable to distinguish between "patently improper" and "properly managed" sites rests on the assumption that regulated, licensed facilities are, in fact, "properly managed." But neither the record of industry nor that of regulators warrants such an assumption. Before regulation, generators routinely, and without the slightest hesitation, employed the least expensive, environmentally most harmful disposal methods. When regulation was finally on the nation's political agenda, they argued that regulators should forbear from interfering in their affairs and should, instead, regulate the disposal industry. They vehemently resisted the idea that they should be made legally liable for their wastes. After the passage of RCRA, generators used every opportunity to try to minimize the impact of the new regulations on them.

Neither does the treatment and disposal industry's record inspire trust and confidence. The Council on Economic Priorities (CEP) has conducted a major study of the eight largest, presumably most capable, corporations in the commercial hazardous waste management sector.[25] As part of this study, CEP examined in depth a sample of ten facilities owned by one of these corporations. The *best* of these had the following enforcement actions taken against it by the Utah Solid and Hazardous Waste Commission in one year, 1984:

18 violations, including (1) improper storage of primary oil/solids/water separation sludge from the petroleum refining industry . . . , (2) failure to manifest shipments of listed hazardous wastes, (3) failure to ship hazardous wastes to an approved TSDF, (4) failure to obtain proper waste analysis, (5) acceptance and management of wastes not specified in the current Part A permit application, (6) exceeding storage capacity at the drum storage dock, (7) failure to segregate incompatible wastes in containers on the drum dock, . . . (9) failure to develop, install, and maintain fire-fighting capabilities. . . . failure to minimize the possibility of any unplanned release of contaminated

pieces of plastic to air, soil, or surface water, placing noncontainerized or bulk waste containing free liquids into the landfill cell. (Goldman et al., 1986:202)

Word limits imposed by my publisher prohibit me from quoting the full list of enforcement actions taken against the *worst* of the ten facilities in the CEP study.[26]

Overall, CEP's evaluation of the top firms in the industry suggests that a host community is not acting at all irrationally when it refuses the nominal distinction between "patently improper" and "properly managed." Experts have shown that even the best-designed landfills are certain to fail.[27] The literature is full of examples of "properly managed" facilities that are malfunctioning, leaking, belching black smoke.[28]

Finally, as I have shown throughout this work, facility permitting and inspection have been so spotty and weak that the public cannot reasonably assume that regulators are protecting them. Citizens cannot trust that the distinction between "patently improper" and "properly managed" sites means anything.

Are People Narrowly Selfish?

I have provided the beginnings of an answer to this charge in chapter 4. People did start out with a narrow NIMBY consciousness, back around 1980. As I have shown, however, the movement now espouses an analysis that is not at all narrow or NIMBYish. Its radical, systemic critique and its vision of economic and political reform are very far from the NIMByism lamented by officials.

Now, it is still true today that when people who have not previously been touched by the toxics problem become involved for the first time, they begin with that narrow, "not here, not me," point of view. However, to the degree that they come into contact with the movement's leading organizations—and I would hazard a guess that most local groups today do make such contact rather early on—they are encouraged to widen their sphere of concern, to think globally while they act locally. I will return one last time to this question of the movement rank and file's subjective understandings in chapter 8.

The Long-Term Societal Rationality of the Shift toward Source Reduction

The final point of the second response to the critics' charges was made in my discussion of the turn toward waste reduction. Environmentalists and scientists, government and industry, now agree that pollution prevention through source reduction is the most desirable, possibly the only workable, solution to the nation's waste problems. I have argued above that it was the phenomenon of decentralized mass refusal that created and continues to create the conditions that make waste reduction an increasingly attractive choice. Even if local resistance results in some increase in unsafe disposal or illegal dumping, in the short run, mass refusal produces a rational result in the long run because it is the force that is pushing society toward a less antagonistic, more sustainable relationship with nature.

Chapter 8

Broader Political Implications?
Environmental Populism and the
Reconstitution of Progressive Politics

The toxics movement produced historically significant innovations in both environmental politics and environmental policy. It achieved an original and innovative synthesis of environmentalism and populism (chapter 4); it was pivotal in the turn toward source reduction (chapter 7). Important as these accomplishments are, I would like to suggest that this movement may yet prove to have political effects that go even further, that transcend the boundaries of its original, single issue.

The movement's ideological development did not stop when it got to radical environmental populism. Its leading organizations have gone on to articulate a much broader social justice perspective that depicts the fight for the environment and for more traditional social causes as two facets of a single struggle. When individuals participate in the movement, they experience what it means to organize, to be an activist; they are exposed to radical perspectives; they are more likely, movement leaders explicitly hope, to identify with, support, and join with others who are fighting other social justice battles.

The Movement's Infrastructure Embraces a
Broad Progressive Agenda

Mainstream environmentalists continue to restrict themselves to their one cause and, at least officially and publicly, tend to avoid becoming involved with other social causes. Perhaps they believe organizations are most effective when they devote themselves exclusively to a single cause; perhaps they fear that they would alienate important segments of their base of support if they were to embrace controversial causes that emphasize injustices based on race, social class, and so on.

The hazardous waste/toxics movement's development has taken it in quite another direction. Its leading organizations explicitly identify the movement with what they depict as a long and proud history of social struggles in America, with the labor movement,[1] the civil rights movement,[2] the antiwar and New Left movements of the 1960s,[3] and the women's movement.[4] The movement's core organizations reach out to and seek common ground with a variety of causes, ranging from older ones, such as labor, race, and women's rights, to more recent ones, such as homelessness and AIDS.

Why or how this evolution from environmental populism to a broader social change agenda happened is not difficult to discern. The key moment was the refusal of the tactical and ideological alternative, mainstream, reform environmentalism, and the articulation of the perspective I have labeled radical environmental populism. Having made that "choice," the toxics movement found itself squarely within a broader left/opposition culture consisting of numerous other social movements that had similar views about the root causes of various oppressions and, also, similar tactics. In addition, and more specifically, the movement found itself necessarily dealing with issues of class, racism, and sexism.

Toxic victims are, typically, poor or working people of modest means. Their environmental problems are inseparable from their economic condition. People are more likely to live near polluted industrial sites if they live in financially strapped communities. Some are exposed to toxics in the workplace, as well as at home. Conversely, communities' and individuals' economic dependence on such facilities can make them vulnerable to job blackmail. The movement has, thus, learned to spell out the connection between economic vulnerability and toxic exposure.[5] It has had to affirm that communities can have economic development without having to accept that there are "necessary" zero-sum trade-offs between jobs and environment.

Issues of race and racism have had to be dealt with because toxics production and disposal takes place to a disproportionate degree in or near communities of the working poor and of people of color. Core movement organizations that wished to reach out and help organize minority communities or poor, multiracial communities had to deal with racial tensions, not just toxics.[6] Thus, in contrast to earlier environmentalisms, where there was little attempt to link that cause to the civil rights movement and other movements against racial inequality,[7] the intersection of race discrimination and exposure to toxic hazards is one of the core themes of the toxics movement.

Conceptually, this intersection of heretofore separate causes has given birth to a new emphasis on "environmental racism" that enriches and deepens both social issues. The struggle for racial justice has been extended to new grounds. Environmentalists are being challenged to rethink their approach to their issue:

> The principle of social justice must be at the heart of any effort aimed at bringing Blacks into the mainstream of environmental organizations in the United States. . . . [We] must not misuse concern for endangered species as a way of diluting our responsibility to meet basic need for human health care, food and shelter. . . . Environmental organizations can no longer afford to take the view that they are unconcerned about who benefits and who loses from restrictions on economic growth. (Anthony, 1990)

In October 1991, the First National People-of-Color Environmental Leadership Summit explicitly redefined *the environment* to include "the totality of life conditions in our communities — air and water, safe jobs for all at decent wages, housing, education, health care, humane prisons, equity, justice" (Southern Organizing Committee for Economic and Social Justice, 1992).

The movement has had to deal with sexism, or patriarchy, because the focus of toxics organizing is home, community, integrity of the family, health — all traditionally women's domain of concerns, and because, as a consequence, women make up the majority, probably the vast majority, of both the membership and the leadership of movement organizations.[8]

In moving from a position in the traditional sexual division of labor to a very different daily life as activists, women had to confront sexism and patriarchy, both as an internalized barrier and as external resistance.

> All of our lives we are taught to believe certain things about ourselves as women, . . . Once we become involved with a toxic waste problem, we need to confront some of our old beliefs. (Zeff et al., 1989:31)

> Many of the struggles and obstacles that these women face as leaders stem from conflicts between their traditional female role in the community and their new role as leader: conflicts with male officials and authorities who have not yet adjusted to these persistent, vocal, and head-strong women challenging the system. . . . Some of the men who are active in our organizations may also treat us the same way. (ibid.:25)

Recognizing that women in the movement must come together to discuss their common experiences and share, give, and get support, CCHW organized a conference titled "Women and Toxics Organizing," where women were encouraged to "think about situations in which you've felt degraded, dismissed, patronized, used, or ignored by men, whether they were government officials, corporate spokesmen, environmental activists, scientists, attorneys, or members of our own groups; . . . share their frustrations and their creative solutions for coping with and for *changing* these patterns of sexual discrimination" (ibid.). Challenging "male chauvinism" was said to be an integral part of building the movement.[9] In this way, the hazardous waste and toxics movement has begun to provide its own answer to ecofeminists' call for an integration of women's liberation and environmentalism.

The result, both of the general place in the political spectrum in which the movement found itself and of the specific organizing challenges it has had to deal with, has been that the movement has, in recent years, affirmed that environmentalism is a social justice issue that must necessarily forge solidarity with all the other great social causes of the day.

> Environmental justice is a people-oriented way of addressing "environmentalism" that adds a vital social, economic and political element. . . . the new Grassroots Environmental Justice Movement seeks common ground with low-income and minority communities, with organized workers, with churches and with all others who stand for freedom and equality. . . . Environmental justice is broader than just preserving the environment. When we fight for environmental justice, we fight for our homes and families and

struggle to end economic, social and political domination by the strong and greedy. (CCHW, *Everyone's Backyard*, 1990, 8[1]:2)

We can reach out to others who, in their own ways, are struggling to recapture lost values of individual rights, fairness, justice and equality of opportunity: activists for civil rights (Asian, African-American, Hispanic, and Native Peoples), organized labor, women, gay men and lesbians, housing and tenants' rights, farmers, welfare rights, affordable health care, animal rights, consumer rights, and safe energy, among others. (Peter Montague, Environmental Research Foundation, 1989:104)

Social guidance of technological decisions is vital not only for environmental quality but for nearly everything else that determines how people live: employment; working conditions; the cost of transportation, energy, food and other necessities of life; and economic growth. And so there is an unbreakable link between the environmental issue and all the other troublesome political issues. . . . environmentalism reaches a common ground with all the other movements ["civil rights, women's rights, gay and lesbian rights, antiwar, against nuclear power and for solar energy, world peace, . . . the much older labor movement"], for each of them also bears a fundamental relation to the choice of production technologies. (Barry Commoner, in *RACHEL*, 30)

Some have even begun to suggest that grass-roots waste and toxics environmentalism is now in the position of leading that larger movement, of being the place where a broader movement for social justice can be reconstituted in future years.

We see this as, this is the civil rights movement of the nineties. This is the antiwar movement of the nineties, and this is an area of tremendous potential social change. And we want to build it. We want to see it happen. . . . The toxics issue is a very powerful, powerful issue to organize people for social change. We feel that it is time to really start being more self-conscious about that. (Gary Cohen, National Toxics Campaign, interview)

At the Level of the Individual: Toxics Activism as Transformative, Radicalizing Experience

Radicalization at the organization level has been matched by radicalization at the personal level. Naturally enough, that process is seen in its most dramatic form in the lives and ideas of the movement's core of leaders. Once again, recall Lois Gibbs's total transformation from apolitical housewife and mother into a dedicated activist, a leader, someone who can articulate a broadly progressive perspective:

Grassroots efforts . . . make America great and bring about change. These efforts are critical for change. Our history has shown us this as far back as the Boston Tea Party. . . . minorities, poor, homeless, workers and leaders of grassroots environmental groups . . . [w]e all have a common bond: victimization. And we all have a common goal: justice for all. ("Together We Can Win Justice," in CCHW, 1989)

Recall, too, the three other leaders who were quoted at length in chapter 4, Sue Greer, Kaye Kiker, and Lew Dunn. All remembered their earlier selves, narrowly focused on their private lives, disinterested or naive or diffident when it came to the public sphere. All described how the process of participation changed them in every way. It transformed their understanding of the world. It changed their self-image, and, in effect, totally changed how they defined their lives. Disillusioned and angered by their experiences, they moved toward a radical critique of society, business, and government. Their new political understandings, their anger, or their deeply felt ethic of responsibility made them, however reluctantly, accept the role of "activist." They have become public persons. In that process, they have not only changed their everyday lives—how they spend their time, what they think about— they have also become more confident, more skilled. They think of themselves as "hardly being the same person" they were before it all began.

Sue Greer expresses her political views:

If our forefathers came back and saw what was happening in this country they'd be shocked and appalled because it's not what their intent was. They have a bunch of people that are bred into corruption, they drink and carouse around, waste our money, they're greedy, they lie, they cheat, they have conflicts of interest. I mean, they're involved in multinational corporations, they cater to them and we are the losers for all those people.

Citizen pressure . . . is the biggest thing we believe in because we think that it's people who have changed this country. Time and time again, changes come from the people. . . . when enough people demand change, things will change.

We organize and we often question the "system." . . . others throughout the history of the United States felt these same feelings. . . . the earlier American radicals who fought for our Bill of Rights, Thomas Paine who fought to abolish slavery in the Declaration of Independence, . . . the Boston Tea Party. Today, we are reaping the benefits of their efforts.[10]

Greer has this to say on personal transformation:

We had a special legislative hearing just for us . . . in Indianapolis. The first time I was ever there in my life, Hal. I was 37 years old. I had never been there. I had never been to Indianapolis. And I went there. I was terrified out of my skin. . . . that was way off my tether. . . . it was hard enough for me to understand all the words, all the new things that I had to deal with, but I had to deal with another problem, that of stepping off of this little circle that I was in and going out into the outer world. And that was real difficult for me. . . . I was traumatized by this whole thing. The fact that I had to step out of a whole secure world, I was very secure, and I had never done the things that I do today. . . . I had to do a lot of growing myself.

Tuesday night bowling, you know, I had to give up all of that. I was a classic league bowler and I had to give it up. And I gave up a lot of things. I gave up TV and everything else to learn.

I like being in the movement. . . . you have an inner satisfaction from helping others and doing good things.

The one thing that's happened since I got into the movement is that now the world seems so small. I know people in foreign countries. . . . To me, when I was first in the movement, and I had only ventured about two hours from my home, the world was so big . . . it scared me to death and now I think it is such a big beautiful world and there's so many people out there. I travel. I stay in people's homes and I feel comfortable with my own [movement] people.

Q: And how are you holding up in this work? Does it sustain you?

A: Yeah, it does. I like it. It's like the nicest thing I have ever done. You know, have you ever gone to work and you hate your job and can't wait 'til 5? You don't do that here. Everybody works here and when we get here nobody is in a hurry to leave. It's a great place to work. . . . You get a whole lot of gratification out of it.

Kaye Kiker of Alabamians for a Clean Environment and the National Toxics Campaign made these comments when interviewed:

Most people are afraid to speak out in America, because of job blackmail, threats to their life, or whatever. When you're fighting multibillion-dollar companies like this one, you put yourself in a lot of danger, really. But, if democracy is in any way useful at all, we're going to use it to benefit this community.

I get calls from all over the United States. I'm getting more and more and more, and it's getting to the point—in mail-outs and things like that—it's impossible to help everybody.

Because of my activity with ACE, I think I could get involved with any kind of issue and be effective. It's helped me in my confidence. I know what I can accomplish. I'm more confident, for sure.

Q: Were you politically involved before you got involved with hazardous waste?

A: Oh, no. Nothing.

Q: What were you up to?

A: I worked . . . I was restoring my home. I was involved in social clubs around here, the Historical Society, the Order of the Eastern Star of Worthy Matrons, Garden Club, Homemaker Club. . . . Church: choir member and assistant Sunday school teacher. . . . I sewed. Made curtains. . . . That was my life.

Q: I see. But nothing that had to do with politics?

A: No, not at all.

Q: Are you still doing that [all the church and social club activities she did before ACE]?

A: No, no. Don't have time . . . out of the state every weekend. . . . Had to get out of it because of the time I spend on environmental issues.

Q: Are you the same person you were five years ago?

A: Oh no, no. Not at all. I can't even remember what I used to think about, to be honest with you. My home was my favorite hobby. Making curtains and decorating tables, having big meals for family, sewing my clothes.

It's been eight years now. I've decided that I'm so far along now that it would be such a waste to just drop what I'm doing. 'Cause we're able to help other communities.

I often think about that: Would I like to be ignorant and unaware again? And I don't think I would. I mean, I did for a long time, believe me. For years, I wished, I kept thinking in the back of my mind that one of these days I'm going to go back and my life is going to be like it used to be. I'll be one of the crowd. But, it wasn't but a few months ago I decided and made a commitment to what I'm doing. It's not what I wanted to do, but it's better to be committed and dedicated than to keep wishing that things weren't the way the were. So, I've sort of decided that I'm not going to expect my life to be like it used to be.

I would like a private life, but that's not, I don't think, ah, well this is not Heaven, so there's always going to be problems. Unfortunately, they seem to get worse, drugs, the economy. . . . there's a lot of big, big problems out there.

Q: If you didn't have to do *this* work, would you be working on something else?

A: Oh, yes, probably would.

Lew Dunn, Greenpeace organizer from Casmalia, California, also went through a transformation:

I [once] thought, " . . . the government will take care of us and won't let people screw us and kill us." And you're naive to believe that. You're just naive to believe that because that is not what's happening in this country today. It's not happening.

I believe we have a corrupt system. . . .

Why don't more than 20 percent of the people vote? . . . Because they don't feel that they count. . . . The democratic process is almost destroyed in this country.

Any revolution or any people that fights back, they fight back when they have nothing to lose and there is nobody listening anymore in government. . . . We are not far away from that unless things get changed.

Q: Before [you got involved with local water issues and the dump], had you been politically involved?

A: No, no.

Q: Not even in voting in elections?

A: No, no. I probably haven't voted in four elections in my life.

Q: And what would you have thought if something went wrong? Like a toxic waste problem, for example?

A: I never thought of it.

Q: And did you think that corporations were favored by government over citizens or that citizens had equal standing?

A: I really never gave it a thought, quite frankly. I never gave it a thought. . . . I had no thoughts about GM, government, or nothing. My main concern was raising my children and making sure that they had food on the table, that they had clothes and that if they wanted to go skin diving that I made enough money for that or if my wife wanted a new car or whatever, we could afford to have a nice car and a good place to live. That was my main concern in life.

I'm a different person. I'm not the same person I was.

It's like a Pandora's box. Once you open it, you don't put it down. Ever. It changes your life forever.

In 1987, I think it is when I went on my first [trip to help a community organize itself]. In Amador County. . . . From there I went to Maricopa, to Avenel, Kettleman, Benicia, Richmond, Martinez, Tehama County, Redding, San Rafael, Newberry Springs, Ludlow, Alpaugh, McFarland, Kern County, Shafter, Susanville, Yerrington (Nevada), Lone Pine, Bishop. . . .

Q: These are in 1987?

A: These are last year [1989]. I've done that with Greenpeace in the last year. That's a lot. I put 70,000 miles on my car.

I get calls from people all over the United States.

Q: Your circuit is?

A: Anywhere they want me. Give me a plane ticket and I'll go anywhere. I am not afraid or intimidated by anybody in government.

I've been on the front lines. I have the confidence in myself.

I show them [community groups] what can happen and what does happen. . . . When they see that the regulators are not interested in regulating as much as they are in making money off of their suffering. I want to outrage them. I want to make them mad. I don't want them to get apathetic ever again in their lives. So we raise a consciousness that they already have. They got the spark; we want to make the fire get going so that these people can defend themselves.

> They thanked me. They said without you coming up and telling us and show-
> ing us that we can beat those people, we would have never got it done. Gosh,
> that's the biggest honor that you can pay me.

> I am pleased that I am able to do it. That makes me feel good.

Such fundamental personal transformations happen to only a small fraction of
the people who become involved in a local toxics cause. Movements are always
characterized by different levels of participation. The hazardous waste and toxics
movement is no exception. In each group, there is a handful of core members who
make the group's work a personal priority, lead the group, keep it together, keep it
going. An even smaller number will be totally swept up into the life of the movement. These are the Lois Gibbses, the Sue Greers, the Kaye Kikers, the Lew Dunns.
Will Collette, the national organizing director of CCHW, estimated how many such
radicalized full-time leaders/activists the movement has created:

> If we look at it from the standpoint of how many people jumped into the issue
> with both feet, in for the duration of their issue, were real rough, tough fight-
> ers, it would number in the tens of thousands. Where you would start seeing
> high percentage drop-offs is: How far does that go? How many people go be-
> yond the life of the issue that got them started? You're looking at, perhaps,
> one out of 200 or one out of 500 will go on to another fight and maybe one
> out of a thousand, or perhaps even higher, will come out of a local fight say-
> ing, "I believe this is going to be my life's work." . . . If pressed, I could prob-
> ably come up with a list of about 100 [names of such folks]. (interview)

If we believe movement figures that, at various times in the past decade, at least
2,000 and perhaps as many as 5,000 local groups have existed, we may reasonably
think that on the order of something like 10,000 or so people found themselves at
an intermediate position in the movement, at the responsible core of a local fight.
Such people have more than a just a fleeting experience in the movement, even if
they do not become permanent, full-time toxics activists.

People at this level are the ones who spend time with the core SMO organizers
helping the local group. They read the newsletters.[11] They are the ones who are
likely to make contact with local activists working on other issues:

> When they talk about how we can change that [when they have concluded
> that "the root of the problem is that we're economically weak"], one of the
> things that comes up is economic development, sound economic develop-
> ment. And as a result of these conversations, they come into contact with the
> local economic development groups that are fighting for affordable housing
> or fighting for tax relief or looking at how to build an economic base. And
> then, going through that network, they often connect with other sorts of net-
> works. Women's rights networks. . . . It's a building thing. . . . then all of a
> sudden they will see the same congressional person who's opposed to choice,
> who's promoting the incinerator or whatever the terrible facility is, . . . or

whatever it was, and they see the connections in a much larger way. (Lois Gibbs, interview)

They attend leadership development conferences or a national convention. There, they get training, they got more exposure to the movement's perspectives, they are brought into active, face-to-face contact with other political causes:

> [At the CCHW convention] in Arlington a couple of years ago, people there were marching on AIDS and we went over and marched with the homeless. . . . People went over to see the AIDS quilt. . . . So [the local toxics activists from] a small rural community see these things happening and realize that they [the homeless, AIDS activists] are people like them. . . . It broadens them; they develop an awareness that all people are the same. (Marty Chestnutt, interview)

> We always do an action at our conventions because we want people to experience an action so they are not fearful of one when they go back. . . . in '89 we decided to join the Housing Now march. . . . we weren't really sure whether our people were willing to cross that, to say, "What does this housing have to do with environment?" and "Why are you asking us to do this silly march?" . . . 600 [toxics] people participated in that march and actually not only saw the crossover but went and talked to folks who were at the march about their common ground, which is oppression, which is money issues. . . . And they saw other things because there was a lot of different groups that supported the Housing Now march. So they were able to pick up the literature and understand, again, the power structure, and how people are oppressed, all different types of people, all different types of issues. (Lois Gibbs, interview)

Such intermediate participants would be expected to experience the personal and ideological changes, the redefinitions of self and society, described above, though to a lesser degree.[12]

Finally, a group's core members mobilize a much larger number of people who sign petitions, contribute money, occasionally come to meetings, people who will come out and take part in a pivotal public hearing or show up at a demonstration. For the groups discussed here, hundreds, sometimes thousands, of citizens are routinely mobilized at this level of participation.

One does not expect to see profound personal or ideological changes in people who are only minimally involved and only for a short time. Still, when people take part in a local action today, one of the movement's infrastructural organizations is usually there to help. When local organizations get that help, when they are told that they are part of a larger movement, they are encouraged to identify with it, stay involved, help others.

> When you block an unsafe facility, shut down or clean up a polluter and win the clean-up of contaminated sites, you add to the nation-wide movement for environmental justice, even if you never leave your own backyard or do another thing! When you share your knowledge and strength with another nearby community, . . . then you're *really* getting into the Movement spirit!

. . . We have a rule at CCHW: "If you expect to get help, you've also got to expect to give it." (Lois Gibbs, in CCHW, *Everyone's Backyard*, 1989, 7[2]:1)

When we work with our groups, we tell them we do not allow a NIMBY syndrome in this organization. I mean, "not in anyone's backyard." We have been very harsh with some groups and said, "Look if you are expecting us to come in here and help your group, and then you get your facility shut down, and you do not go out and help others, you can count us out right now. 'Cause our goal is to multiply you." (Sue Greer, interview)

[Take] Marion County, Tennessee, for example. They tried to put a hazardous waste incinerator in there and they defeated it; they ran them out. And when they called me and told me they had won, I said, "You haven't won until you find out where they're planning to put that incinerator and you help the next group," and within about a week they called me back, said, "We found out where it's going. We've already been in contact with the people over there and we're helping them organize." (Marty Chestnutt, interview)

The groups being helped are also given the movement's literature, told about a much larger environmental crisis, exposed to the movement's latest explanation of the causes of and cures for environmental problems. They are exposed, too, to the movement's increasingly explicit identification with other causes. Ideological development at the level of the movement as a whole means, then, that even at the most fleeting level of participation, citizens who take part in local actions get exposed to alternative, radical explanations of their social world.

Of course, in spite of all encouraging and cajoling, most people's participation will be limited and of short duration. Lois Gibbs has noted that, "for the vast majority of groups in the Movement, the local fight is everything" (in Zeff et al., 1989:39). Sally Teets, Midwest organizer for CCHW, said in an interview for this volume, "People that see themselves in their fight often don't want to join a larger group [i.e., state coalitions] where there are more meetings and fight because it becomes overwhelming." And even as Sue Greer talks of being "very harsh with some groups" because they do not want to keep fighting, to help others, she notes, "it does happen a lot [that they quit] because people get tired of working on the issue" (interview). As Marty Chestnutt put it, "Most groups are one-issue groups. Once they win a case, they fold up and go away. One or two people will want to continue, get fired up" (interview).

Many will leave with the same narrow NIMBY consciousness they started with. Even so, leaders believe, something will have happened in people's understanding of their world, their willingness to experience solidarity with others, their willingness to fight again another day:

When you first go in you think that you are the only community, the only one this is happening to. As you reach out and finally connect with a national group and you start getting their newsletters and you start reading what they are saying . . . you get that larger connection. You start seeing the bigger picture . . . broaden your horizon. . . . A lot of these people that I worked with were very narrow-minded, extremely conservative: "There's only one right way and it's the way my church . . . " I've seen a lot of those people expand and start to realize that they don't live on an island. (Marty Chestnutt, interview)

Gibbs says that that is okay, that it doesn't matter if a group folds after they win their fight. Because people have learned how to fight and win. They will know how to fight. They will go on to other things even if it isn't environmental. It might be women's rights, or working for minorities or it might be gay rights. But they will be quicker to fight oppression because they have been oppressed. (ibid.)

Chapter 9

Concluding Remarks

Sociologists, unlike our colleagues the historians, tend not to feel that our work is done when we have composed and presented a concrete, distinctive narrative; typically, we cannot rest until we have wrested some more abstract, generalizable points from that narrative. In that spirit, I wish, first, to briefly indicate what are some more general methodological, conceptual, and political lessons that we may draw from the history I have presented in this work.

"Issue History" as Methodological Approach

Studies in environmental sociology or political sociology tend to privilege one or another individual zone of society's total field of political practices, be that people's perceptions (media studies, public opinion, risk perception), popular action at the base (social movements), or formal policy (policy studies, studies of regulation). In doing this research, I found it impossible to confine myself to any single subfield of inquiry. "Hazardous waste" exists as representation in mass media and as a perceptual object in the popular imagination. Hazardous waste is the object of social movement organizing. Hazardous waste is addressed by a multitude of state and federal laws. To reconstruct the history of hazardous waste, I had to be open to following the issue as it moved from one sector of social process to another, to move my gaze from regulation, to perceptual processes of mass opinion formation, to movement organization, and back again to the formal side of the political process. I had to learn to be flexible and eclectic, to move as needed among the various relevant sociological subfields. I have come to think of this approach as the study of *issue history*.

I found that doing issue history requires both a heuristic and a synthesizing approach to theory and methods. Following the issue's development required, I found, that I make use of a wide array of data materials. For chapter 2, in which I recount the passage of RCRA, the first toxic waste law, for example, I had to examine government documents, EPA reports, minutes of congressional hearings, and the like, in great detail. For chapter 3, on the formation of mass beliefs about "toxic waste," I had to analyze videotapes of television news, inspect articles in popular magazines, review opinion polls, and read still more government documents. For chapter

4, the chapter on the grass-roots movement, I inspected movement publications and transcribed in-depth interviews with movement activists.

No single theoretical framework could, by itself, either guide the investigation or help interpret the materials that I assembled. To make sense of the materials in chapter 2, I found it necessary to bring to bear several different theoretical strands, from Polanyi's economic history and neo-Marxist state theory to regulatory studies in mainstream political science. Interpreting the media stories and their impact on people's perceptions, in chapter 3, required the help of structuralism, semiotics, discourse analysis, and theories of postmodernity, as well as a host of more "local" theories developed within the subfields that study media, risk perception, and public opinion. So it went with every new chapter in the issue history. The issue-history approach required a nondogmatic, eclectic attitude toward theory, a willingness to bring to bear whatever proved helpful for reconstructing the full history and for analyzing or interpreting its meanings.

Conceptually: Attending to the Transitions and Interactions

Let me now make a distinction between moments when events unfolded predominantly *within* a single zone of politics—in the policymaking bodies of government, say, or in media-driven opinion formation, or in the grass-roots mobilization of communities—and those moments when the locus of events moved from one zone to another or when things happened in several zones simultaneously. It seems to me that one lesson to draw from this issue history is that the dynamics of action in any one zone or arena of politics are generally well understood and are not, by themselves, remarkably new or interesting; in contrast, things become conceptually interesting exactly when action transcends the boundaries of any one zone. The challenge and the conceptual richness lies in identifying the conditions that allow action to jump from one zone to another and in describing the synergies that then arise.

For example, the dynamics of regulatory law formation and evolution have been exhaustively studied and are well understood. That Congress would have sleep-walked into regulating a new area of industrial activity is not likely to strike students of the legislative process as something remarkable. But it is quite interesting to study how, once public perception of "toxic waste" changed, the flaws in what Congress had wrought generated settings and occasions for grass-roots mobilization.

Similarly, risk perception theory, social problems theory, and studies of postmodernity, critically combined, provide a persuasive reading of the process of contemporary, media-centered issue formation. The way that media coverage produced mass perceptual change was, by itself, not particularly interesting. The conceptual challenge came in understanding the conditions that allowed those attitudes to become the basis for the emergence of a new social movement.

Likewise, the dynamics of community-based mobilization have been explored with great care in the social movements literature. All the talk of "new" social movements notwithstanding, the notable thing about the hazardous waste and toxics movement is that it was largely a rediscovery of a quite traditional form of popular, grass-roots organizing. The only thing "new" about it, really, was the issue that sparked it off. The motivations to act were very straightforward, all responses to perceived threats to what one has or thinks one is entitled to, be that "material" things such as health and property or "ideal," moral-economic things such as fairness. The process of organizing was also straightforward and unremarkable, a rediscovery of long-familiar tactics and rhetoric. The processes through which participation produced increasing radicalization are also familiar ones. Conceptually, what was most interesting about this social movement was not its predictable course, but the way it provoked contradictory responses and how those contradictory responses produced novel policy events.

Politically: The Synergies of Multiple, Simultaneous Tactics

Still in the spirit of generalizing, we might draw from our story a political lesson that parallels the methodological/conceptual one. That is, political activities that are confined to a single zone tend to yield disappointing results; in contrast, when political events occur simultaneously in several different zones, the interactions that ensue among them tend to generate real forward motion.

Action that occurs in isolation, in a single zone only, is not likely to produce significant successes. Studies of regulatory failure amply demonstrate that political action that depends exclusively on securing passage of regulatory laws is certain to fall far short of its goals. Postmodern issue creation may produce quick attitudinal impact but, if nothing further happens, the quality and staying power, and therefore the political meaningfulness, of those attitude changes are suspect. The presence of grass-roots action is always a key to real forward movement, but, as this case suggests, it is difficult to mobilize communities if some more general attitude-formation process has not yet occurred, and it is more difficult for movements to define targets and tactics if there are no failing statutes to organize against, no procedural safeguards to (ab)use.

Things happened politically when actions leapt from one zone to another and activities in several zones, going on simultaneously, had impacts on one another. Once the media had done their work and produced a new issue-icon, RCRA could no longer just sink quietly into the familiar state of chronic, but largely ignored and invisible, failure. Postmodern issue creation became meaningful when its attitudinal effects helped transform isolated, sporadic local actions into a true movement; conversely, the movement reinforced and stabilized the attitudes fostered by the icon. And, as I argued in my discussion of source reduction, in chapter 7, policy evolution took innovative directions when popular organizing interacted with regulatory law in complex and contradictory ways.

Historical Specificity and Significance

Such reflections suggest that the task, both intellectually and politically, is not just to understand what happens in each zone of activity, but more to understand the conditions under which issues jump from zone to zone, creating complex, dynamic interactions among them. Perhaps, however, it is better to resist the sociologist's tendency to boil the concrete down into abstract, ahistorical generalizations and choose, instead, to appreciate this history's specific meanings, its historical accomplishments.

This line of reflection begins with the observation that this issue history took place not only at a time of growing concern with environmental crisis, but also at a time of two unrelated political crisis moments. One of these was specific to environmental politics. It concerned the growing realization that the cornerstone of modern environmental politics, the struggle for stronger regulatory law, had produced, at best, equivocal results. The other concerned progressive politics more broadly and had to do with the decline of the labor movement, hence its inability to continue to be a leading force for social change, and the New Left's failure to sustain its attempt to provide the leadership that had thus been lost. As the hazardous waste movement became, first, the "toxics movement" and then the "movement for environmental justice," it seemed to provide a response to both these crises, certainly the former, possibly the latter.

The hazardous waste movement proved a major intervention in the development of environmental politics. The movement articulated an innovative new perspective, radical environmental populism. It brought whole new segments of the population to environmentalism. It broadened environmentalism's tactical choices by demonstrating how much can be accomplished through a politics of direct, grass-roots organizing.

The movement helped put source reduction or pollution prevention on society's agenda. Social scientists had long argued that the standard logic of regulation does not work. Environmentalists, after two decades of struggling to pass and implement traditional regulations, have begun to share that conclusion. Increasingly, waste reduction is said to be the workable alternative. It is the toxics movement, however, that not only talked waste reduction but created and continues to create the conditions that compel actual movement in that direction. This is an important and timely accomplishment.

Is grass-roots activism *the* answer to the pressing question of how we are to deal with the environmental crisis of our time? Of course not. Many of the environmental problems we face are the result of complex interactions among patterns of consumption and population as well as production. Still others, notably global atmospheric changes, do not link to immediate experience in any obvious and compelling manner. Community-based toxics activism is not well suited to addressing such problems. On the other hand, the grass-roots movement *has* made a contribution toward dealing with other environmental problems in the important sense that it has expanded environmentalism's demographic base and has taught people who were initially concerned only with a local, immediate, directly experienced threat to see their plight as part of a much larger, systemic problem.

Important as those accomplishments have been, the developments described in chapter 8 suggest that the movement might prove to have more general political impacts, still. Movements take on greater historical significance when they move from the particular to the universal, when they expand out from their specific issues of origin and embrace a more global social change agenda. They take on greater historical significance when they not only mobilize participants to fight for their own interests but also provide a broader radicalizing experience. The labor movement, to take the most important modern case, certainly would have had important impacts on historical development if it had fought only for its sectional interests, for shorter hours, higher wages, better working conditions; it took on truly historical dimensions when it embraced socialism and took it upon itself to be the brains, heart, and muscle of a more global challenge to the status quo.

With labor in decline, it has not been clear which, if any, of the "new social movements" would make the move, would go beyond its particularistic cause and take it upon itself to fill the vacated position of leading/unifying element of all the various movements that challenge the existing organization of society. The hazardous waste movement is making explicit gestures in this direction. It increasingly defines its environmental mission in terms of a larger critique of society; it makes common cause with other movements and says that, ultimately, they are all joined in the same struggle. It even envisions a future in which grass-roots environmentalism spearheads the reconstitution of a broad social justice movement.

Will the movement realize its larger political intentions? We cannot know. We do know that the icon persists. People dread hazardous waste as much as they did in 1980. We know the "NIMBY" phenomenon—I put it in quotes to emphasize again what a mischaracterization the term now is—is as strong as ever. The movement continues to evolve ideologically. It has produced a cadre of seasoned, capable, committed movement leaders and introduced many thousands of people to the experience of activism. In a time when movements for social change are at low ebb and radicalizing experiences are rare, *any* movement that is vital and dynamic, that provides opportunities for activism, collective behavior, and exposure to a critical perspective on society is a historically significant event.

Beyond that, all we can say is that local, community-based environmental conflict *may* end up serving the same historical function that factory, shop-floor conflict served a century ago. If so, then its humble beginnings as local NIMBYism will not only have produced an innovative environmental populism, but will have contributed to the reconstitution of a broader progressive movement in the United States. At that point, the boundary of issue history will blur and the specific history of hazardous waste will open up to the larger political history of its time.

Notes

1. Introduction: Environmental Crisis and the Search for a Politics That Works

1. This list was assembled from an examination of the Vanderbilt Television News Archive's *Television News Index and Abstracts* for 1989. That year was chosen simply because it was the last full calendar year for which the index was complete at the time of the research. In retrospect, perhaps 1989 was not the best choice. It was hardly a typical or average year. Coverage was monopolized by one event, the *Exxon Valdez* oil spill in Alaska, which accounted for almost a third of the year's coverage. Coverage of the *Exxon Valdez* may have had a sensitizing effect, increasing overall media attention to environmental problems that year; conversely, in taking a huge share of the news hole, it may have cut into the variety of environmental stories that made it to the air.

2. There were stories about Russian nuclear problems, as well: several stories on Chernobyl, reports about a massive nuclear accident that had been hushed up for decades, and an accident aboard a nuclear submarine. Perhaps this is the place to note that what passed for "actual existing socialism" in Eastern Europe and Asia was, if anything, even more harmful to the environment than capitalism. That fact can be interpreted as supporting the view that the problem is not capitalism but something fundamental to *modernity*, more generally. This is an argument found, in various guises, in Bookchin's social ecology, in deep ecology, in some variants of ecofeminism, and in critical theory (Horkheimer and Adorno, 1987). One could, alternately, make a specific structural and historical analysis of "actual existing socialism" as a distinct mode of production that has its own characteristic mode of destructive interaction with nature. See Gille, 1992. Although I sympathize with some versions of the critique of modernity, overall, I prefer the latter approach. In any case, consideration of the former Second World is beyond the scope of this work.

3. Indeed, Harvard biologist E. O. Wilson declared that the world was in the midst of the worst species die-off since the Mesozoic era, sixty-five million years ago (Shabecoff, 1989).

4. Journalists, concerned physicians, and government inspectors are cited by Marx, 1867/1967, Engels, 1845. As anyone who has read Durkheim's analyses of forced and anomic forms of the division of labor, or even just Dickens's novels, knows, one did not have to be a socialist to be concerned about these impacts. The political economists' observations are reviewed by Polanyi, 1944.

5. See Polanyi, 1944, on effects on both land and labor. Marx, 1867/1967, and Engels, 1845, were primarily concerned with working conditions, but see Perelman, 1990, for a summary of Marx's views on soil erosion.

6. Shabecoff, 1990a.

7. *RACHEL's Hazardous Waste News* is a publication of the Environmental Research Foundation. Hereafter cited as *RACHEL*.

2. Routine Regulatory Failure: The Resource Conservation and Recovery Act of 1976

1. EPA Toxics Release Inventory (TRI) for 1989; see EPA, 1991. Large as this figure is, it

is an underestimate. Not all firms have to report. In 1989, for instance, only 22,569 firms filed TRI reports with the EPA; firms do not have to report releases of less than 10,000 pounds and, in some cases, less than 25,000 pounds; the 325 substances that have to be reported fall far short of the 800 toxics regulated by the state of California. For American Chemical Society, see *RACHEL*, 148.

2. P.L. 94-580, 90 Stat. 2795.

3. Kovacs and Klucsik, 1977:2120; GAO, 1979:1, 1987c:23; House of Representatives, 1987:102; *RACHEL*, 148. After more than a decade of regulation, the General Accounting Office complained (1987c) that the federal government *still* did not have reliable figures on generation.

4. Slovic, Fischhoff, and Lichtenstein, 1985.

5. P.L. 91-512, 84 Stat. 1227.

6. Kovacs and Klucsik, 1977:254-255, 217; Lieber, 1983:62-63; Worobec, 1980:634.

7. The discussion in this section is based primarily on House of Representatives, 1974b, and Senate, 1974.

8. It is interesting to note that the environmental lobby also talked about waste reduction only in connection with solid waste disposal. See testimony from Environmental Action, Friends of the Earth, Sierra Club, and the Oregon Environmental Council, in both House of Representatives, 1974b, and Senate, 1974, part 2.

9. House of Representatives, 1975.

10. Tony Mazzocchi, testifying for the Oil, Chemical and Atomic Workers, was the only exception to the anti-product-control line taken by union representatives in these hearings.

11. A methodological note: Here and elsewhere in the book, I reproduce blocks of quotes in which many voices are, seemingly independently, singing the same notes. Whenever I do this, the purpose is to document the presence of some strand of Official Discourse.

12. These quotes provide only a sample. There were many similar statements from firms such as Monsanto, Alcoa, and Exxon, and from trade associations such as the Texas Chemical Council.

13. Again, the same point was made in these hearings by others, including the New Jersey Manufacturers Association, Shell Oil, the American Petroleum Institute, Alcoa, American Cyanamid, Marathon Oil, and Chemagro Mobay.

14. Piasecki and Davis, 1987.

15. See comments by two EPA officials: Alvin Alm, 1984:1-2, and Thomas Jorling, in Wolf, 1980:491.

16. An excellent description of this struggle over facility financing can be found in Kovacs and Klucsik, 1977:257-258.

17. Again, Kovacs and Klucsik, 1977, provide an excellent review of the final stages of the congressional process.

18. Both policy scientists and government officials frequently cite both "complexity" and "uncertainty" when explaining the EPA's failure to implement RCRA in an effective and timely manner. See Carnes, 1982; Lester, 1983; Lieber, 1983; GAO, 1978a.

19. House of Representatives, 1977, 1978a; Senate, 1978.

20. Szasz, 1982, ch. 5.

21. Others have argued that simple mismanagement also played a role in the Carter administration's failure to get RCRA implementation off to a timely start. See Carnes, 1982; Lieber, 1983:71.

22. Senate, 1983, vol. 2:52.

23. The administration also reined in EPA staffers who wanted to use Section 7003, the "imminent hazard" provision of the act, to prosecute polluters aggressively (Epstein et al., 1982:229-230).

24. GAO, 1979:i.

25. New Jersey, for example, had promulgated a manifest system in 1978 to control "special hazardous wastes" from "cradle to grave," but congressional testimony revealed that until 1980 New Jersey did not have a single person assigned to monitor the manifests being filed in Trenton (House of Representatives, 1981b:124). The following illuminating exchange took place in a congressional hearing: Albert Gore, Jr.: "What enforcement efforts are you making to prevent the abuse of the manifest system?" Edwin Stier (New Jersey Division of Criminal Justice): "The only way the manifest system is going to be properly, effectively enforced is through the proper analysis of the information that comes from the manifest. . . . Anyone who assumes that a manifest system which looks good on paper can control the flow and disposition of toxic waste without the kind of support both technical and manpower support that is necessary to make it effective, I think, is deluding himself. [However,] . . . we aren't looking specifically for manifest case violations. We aren't pulling every manifest in that is filed with the department of environmental protection and looking for falsification of manifests specifically because we don't have the time, the resources, or the specific lead information to do that" (House of Representatives, 1980:140).

26. Friedland, 1981:100.

27. GAO, 1978a, 1978b, 1979. The GAO observed, for example: "The Environmental Protection Agency has been unable to obtain the funding authorized for carrying out hazardous waste disposal programs" (1979:i). "Key activities . . . [including] (1) developing criteria for sanitary landfills and (2) publishing within 1 year an inventory of all disposal sites not in compliance" would not be completed within legislated time frames (1978b:ii). Delays were having "an adverse effect on creation of additional . . . facilities" (1978a:4). "Federal and State agencies have not assessed the extent of damage to groundwater supplies or determined the number of disposal sites which may be leaching" (1978b:i). "Virtually nothing is known about the over 100,000 industrial waste land disposal sites" (1978b:ii). "The financial and technical assistance promised to the States has not been provided" (1979:i). "Most States did not know the volumes of hazardous waste being produced in their jurisdictions and had virtually no information as to how they were being disposed of" (ibid.). "States lack the staff and funds to effectively carry out [RCRA's] hazardous waste requirements" (ibid.).

28. The House Commerce Committee's Subcommittee on Oversight and Investigations took the lead with thirteen days of hearings on hazardous waste over the year following October 1978 (Worobec, 1980:635). Other subcommittees followed suit. See House of Representatives, 1978b, 1979a, 1979b, 1979c, 1979d; Senate, 1979, 1980.

29. Costle and Beck, 1980:432; Senate, 1983, vol.1:60; vol. 3:25.

30. The literature on regulation distinguishes between "economic" and "social" regulation. See Lilley and Miller, 1977; Mitnick, 1980a, 1980b; Szasz, 1984, 1986c. "Economic" regulations, such as of the airline industry, trucking, and banking, regulate relations among firms in the same sector of the economy. They tend to reduce competition among firms by setting prices and restricting entry into the sector. They reduce risks and create a more predictable and controllable market environment. Often, business owners desire "economic" regulation and, once it is in place, strategically manipulate it in their own interest (Bernstein, 1955; Owen and Braeutigam, 1978; Stigler, 1975). "Social" regulation, on the other hand, is imposed upon business by the state, usually in response to popular demand that something be done about the adverse impacts of industrial production. Social regulations threaten, at least theoretically, firms' cherished and jealously guarded free space to act as they please in the realm of production and investment decisions. Firms tend to resist, rather than desire, the imposition of social regulation; once it is in place, they seek to minimize its impact. Thus, though they share a common label, economic and social regulation are fundamentally different types of policies, and they have profoundly different dynamics. It should be kept in mind that the discussion below deals only with social regulation.

31. Stevens, 1990.

32. See Polanyi, 1944. Polanyi makes this point in order to argue that, far from being an inevitable and necessary corollary of human nature, the economics of self-regulating markets was a radical departure from past social practices—a departure, he would further argue, that was likely to produce disastrous results.

33. I must hasten to note that capital's professed yearning to be free of all social restraint, and specifically of state intervention, was, in practice, applied selectively and existed in pure form only in the realm of ideology. As economic historian Maurice Dobb has said, "Capitalism has sometimes been represented as constantly striving towards economic freedom, . . . Capitalism . . . [as a] historical enemy of legal restraint . . . bears little resemblance to the true picture. . . . [Capital] is not at all averse to the acceptance of economic privileges and State regulation of trade in it own interests" (1963:25).

34. See Engels, 1845, and the historical sections, especially sections on the working day, in Marx, 1867/1967.

35. For a history of English hours legislation, see Marx, 1867/1967. American labor historians' work on nineteenth-century movements for the ten-hour day and, later, for eight-hour legislation suggests that hours legislation was an important precedent for regulatory policy in the United States as well.

36. Ashford, 1976.

37. These failings have been documented extensively for the case of worker safety and health legislation. This is, in part, because the labor movement is older than the environmental movement, hence the state has been involved for much longer in statutorily mediating the conflict between labor and capital than the conflict between environmentalists and capital. Therefore, in subsequent notes, I support the various points made about regulation with citations to the literature on worker safety and health.

38. That is, in theoretical language familiar to sociologists, it has to be "constructed as a social problem" (Schneider, 1985; Spector and Kitsuse, 1987). For a parallel effort to conceptualize this process in political science, see Cobb and Elder, 1983.

39. Ashford, 1976; Berman, 1978; Brodeur, 1973; Nugent, 1987; Rosner and Markowitz, 1987a.

40. At Love Canal, residents eventually concluded that there was a connection, and some researchers agreed (Gibbs, 1982; Goldman, Paigen, Magnant, and Highland, 1985; Paigen, Goldman, Highland, Magnant, and Steegman, 1985), but the connection was vague enough to allow others to contest the existence of health impacts (Kolata, 1980; Maugh, 1982). For discussion of continuing uncertainties about the health effects of hazardous waste dumps, see Grisham, 1986; Lowrance, 1981.

41. For the case of worker safety and health, see Berman, 1978; Rosner and Markowitz, 1987b; Szasz, 1982.

42. The American labor movement's fight for regulation of unhealthy factory conditions shows that progress does, indeed, sometimes have to be measured by the decade, if not the century (Berman, 1978; Szasz, 1982, ch. 3).

43. Marxist state theory has articulated three propositions in the analysis of social regulation: (1) States have the *capacity* to adopt social legislation because they are "relatively autonomous," and so relatively free from the short-run interests of and control by capital. (2) They are *motivated* to do so when conditions threaten to bring into question the legitimacy of the social order. (3) However, this policy response is always within strict *limits*. See Block, 1984; Habermas, 1975; Miliband, 1969; O'Connor, 1973; Offe, 1974, 1975; Poulantzas, 1975; Wright, 1979.

44. See also similar comments by Miliband, 1969:77-78; Poulantzas, 1975:191.

45. Lane, 1966.

46. Among most politicians, such levels of intervention are not just not talked about, they are simply unimaginable. As Commoner says, such measures are "so foreign to what passes for our national ideology that even to mention it violates a deep-seated taboo" (1989:12).

47. There is room to struggle over many specifics: Who will set standards, using what criteria? What enforcement powers will the agency have? Who will be exempt from regulation? What kind of budget will the agency have? What kinds of procedural and judicial safeguards will regulated firms have? For the case of worker safety and health, see Szasz, 1982.

48. Downs, 1967; Lindblom, 1959.

49. For the infrastructural vacuum that plagued efforts to regulate worker safety and health, see Szasz, 1982. For the case of hazardous waste, see above.

50. It is estimated that perhaps 12,000 chemicals used in the workplace may be toxic. OSHA managed to promulgate barely ten new health standards in its first ten years of operation. In any one year, OSHA inspected only a small fraction of the more than four million places of work covered by the act and the overwhelming majority of those inspections were for minor, nonserious safety violations. See Szasz, 1982. As I show in this work, RCRA and Superfund have both experienced similar patterns of extremely partial implementation.

51. Bernstein, 1955; Stigler, 1975.

52. Ethridge, 1987.

53. For examples from the case of worker safety and health, see Szasz, 1982, 1984. For RCRA, see Epstein et al., 1982; EPA, 1976; see also the quotes, above, about the millions of pages of comments on proposed standards that were submitted by industry. As I commented in an earlier work, business's willingness to use procedural safeguards in an ongoing effort to weaken regulations is both "comprehensive and reveals exquisite attention to detail" (Szasz, 1986b:19n).

54. For deregulation, generally, see Claybrook, 1984; Grozuczak, 1982; Lash, Gillman, and Sheridan, 1984; Pertschuk, 1982; Sheoin, 1985; Simon, 1983. For OSHA, Szasz, 1982, 1986c. For hazardous waste, Fortuna and Lennett, 1987; Harris, Want, and Ward, 1987; National Wildlife Federation, 1982; Szasz, 1986a; and chapter 6 of this volume.

55. House of Representatives, 1974b:262-276; 1975:263-270, 287-310; Senate, 1974, part 1:21-24; part 2:679-687.

56. See, also, similar comments by EPA staff in the Bureau of National Affairs publication *Environment Reporter*, 1979, 9:2114, 2295; Senate, 1978:15.

57. EPA, 1979:10-11.

58. As early as 1978, Representative Florio exclaimed that "there is a whole new bootleg industry taking this stuff and legitimate manufacturers are selling it to someone who pulls up to the gate in a truck, and they are taking it and dumping it in one of the local lakes or out in the woods" (House of Representatives, 1978a:53-54).

59. Szasz, 1986b. Reports of mob involvement continued well into the 1980s. See Blumenthal, 1983a, 1983b; Oreskes, 1984.

3. "Toxic Waste" as Icon: A New Mass Issue Is Born

1. Moloch and Lester, 1975.

2. Dunlap, 1986, 1987; Mitchell, 1984a; Schoenfeld, Meier, and Griffin, 1979.

3. Based on examination of Vanderbilt Television News Archive, *Television News Index and Abstracts*, 1970 to 1980.

4. Downs, 1972.

5. Converse, Dotson, Hoag, and McGee, 1980; Gallup, 1972, 1978, 1979, 1980, 1981.

6. Dunlap, 1986, 1987.

7. See chapter 6.

8. Gladwin, 1987.

9. Some early cases are described by Epstein et al., 1982.

10. Popper, 1987a, 1987b.

11. GAO, 1978a:11; EPA, 1979.

12. Bennett, 1988; Gans, 1979.

13. Brown, 1979; Epstein et al., 1982; Fowlkes and Miller, 1982; Freudenberg, 1984; Gibbs, 1982; Levine, 1982; Senate, 1983, vol. 1:700-704.

14. Moloch and Lester, 1975.

15. Vanderbilt Television News Archive, *Index*.

16. Levine, 1982:100, 156, 171, 192, 207.

17. The networks gave the story different levels of coverage. CBS devoted the most time, about 83 minutes, ABC about 63 minutes, NBC about 47. Although the sheer difference in amounts of time given to the story would suggest that viewers devoted to David Brinkley and Jessica Savitch would have gotten a somewhat different impression of the importance of the story than, say, viewers who preferred Walter Cronkite and Roger Mudd, inspection of the stories shows that the coverage was so repetitive and so highly stylized that the visual and cognitive contents were essentially identical from one network to another. I selected the full set of CBS stories and a smaller number from the other networks. Following is a key for identifying the stories referred to in the text:

CBS: 8/2/78, C1; 8/7/78, C2; 8/8/78, C3; 8/9/78, C4; 9/1/78, C5; 9/6/78, C6; 11/21/78, C7; 12/14/78, C8; 1/5/79, C9; 1/10/79, C10; 2/6/79, C11; 2/9/79, C12; 2/16/79, C13; 3/2/79, C14; 4/10/79, C15; 4/17/79, C16; 4/30/79, C17; 4/30/79, C18; 5/8/79, C19; 6/13/79, C20; 12/20/79, C21; 2/26/80, C22; 3/22/80, C23; 4/28/80, C24; 5/5/80, C25; 5/17/80, C26; 5/20/80, C27; 5/21/80, C28; 5/29/80, C29; 6/6/80, C30; 8/24/ 80, C31; 9/19/80, C32; 9/29/ 80, C33; 10/1/80, C34; 11/18/80, C35.

ABC: 8/2/78, A1; 8/4/78, A2; 8/7/78, A3; 8/11/78, A4; 9/8/78, A5; 11/21/78, A6; 5/17/ 80, A7; 5/21/80, A8; 9/11/80, A9; 11/20/80, A10.

NBC: 8/9/78, N1; 9/24/78, N2; 4/22/80, N3; 5/17/80, N4; 5/18/80, N5; 5/21/80, N6.

18. In addition to the obvious choices of *Time* and *Newsweek*, I selected the women's magazines because women have been central to the hazardous waste movement. The articles examined are as follows. *Newsweek*: Beck, Lord, and Buckley, 1979; Beck and Lord, 1979; Begley, 1980; Clark, Hager, Shapiro, and Marbach, 1980; Gwynne, Whitaker, Shannon, Hager, and Begley, 1978; Morganthau and Hager, 1980; Sheilds, Cook, Hager, and Carey, 1980. *Time*: Anonymous, *Time*, 1978, 1980a, 1980b, 1980c; Magnuson, Stoler, and Nash, 1980. *Glamour*: Anonymous, *Glamour*, 1980a, 1980b. *McCall's*: Eckman, 1980. *Redbook*: Gallagher, 1979. *Mechanix Illustrated*: Weiss, 1980.

19. Any one picture could, by itself, warrant a detailed semiotic investigation. Look at the woman and child behind the "give me liberty" sign on page 25 of *Newsweek*, August 21, 1978. The overt message is clear enough. The picture is full of class markers, the size and quality of the home, the mother's clothes, both mother's and child's disheveled personal appearance, the uneven block lettering of the sign, the cardboard it is written on. The threat to every aspect of the white working-class American Dream is efficiently signified. Look at our plight. We had a nice little home, white picket fence. family. But look at us now. We're being displaced (in the background, a man is moving boxes of possessions to the porch). My family has been made sick ("I've got death"). Society, help us ("Give me liberty"). In a speculative mood, we might venture that resonances of traditional Christian iconography—the Madonna and Child, now contaminated—give the picture a certain extra unconscious *frisson*; more overtly, the sign claims an identification with the American Revolution, evoking a patriotic slogan that all American schoolchildren recognize.

20. For an overview of the topics of interest in congressional hearings in the 95th Congress, see Senate, 1983, vol. 3:365-369.

21. For summaries of the Commerce Subcommittee's charges, see Carnes, 1982:41; Worobec, 1980:635-636.

22. Levine, 1982.

23. By President Carter, in Senate, 1983, vol. 3:25; by Thomas Jorling, EPA administrator, in Senate, 1983, vol. 1:60; in the Senate Committee Report to accompany S. 1480, ibid.:315; by individual senators and members of Congress, in ibid.: 59, 694; vol. 2:226.

24. For the Valley of the Drums, for example, see Senate, 1983, vol. 1:48, 60, 311, 684; vol. 2:49, 226.

25. For similar coverage in print, see *Newsweek* stories by Beck and Lord, 1979:51; Beck et al., 1979:41; Gwynne et al., 1978:25; Morganthau and Hager, 1980:35; Sheilds et al., 1980:82.

26. Cohen and Tipermas, 1983; Mazur, 1984a; Szasz, 1982:350.

27. Bush, 1981:536; Senate, 1983, vol. 1:708-709.

28. Mitchell, 1984a, 1984b; U.S. Council on Environmental Quality, 1980.

29. Anonymous, *Glamour*, 1980a, 1980b.

30. See remarks by Congressman Volkmer, Senate, 1983, vol. 2:265; Senator Stafford, vol. 1:691; and Senator Tsongas, ibid.:708-709.

31. Clymer, 1989.

32. Slovic, Fischhoff, and Lichtenstein, 1980; Tversky and Kahneman, 1982.

33. Freudenburg and Baxter, 1983; Mazur, 1981, 1984a, 1984b; Rankin, Nealey, and Melber, 1984.

34. For some recent sociological critiques of the psychological approach to risk, see Heimer, 1988; Short, 1984; Szasz, 1988.

35. Regarding sociology, see Spector and Kitsuse, 1987; concerning political science, see Cobb and Elder, 1983.

36. Gamson and Modigliani, 1989.

37. The literature on postmodernity is, by now, voluminous. This discussion is based mostly on the work of Jameson, 1984, 1991; Lash, 1990; Harvey, 1989; Ross, 1988; Baudrillard, 1983; Edelman, 1988.

38. Anyone familiar with critical theory (Horkheimer and Adorno, 1987; Marcuse, 1964), the situationists (Debord, 1977), or earlier studies of the connection between economic growth and advertising-driven production of demand (Baran, 1957; Baran and Sweezy, 1966; Ewen, 1976) knows that some of what is labeled new and postmodern has been with us for quite some time. Similarly, political scientist Murray Edelman has argued, in a series of works dating from 1964 (1964, 1971, 1977), that American politics has long been dominated by symbolic gestures rather than substance. Darnovsky (1990) shows that image management and other public relations methods derived from advertising found their way into American politics as far back as the Wilson administration. What are said to be defining features of "postmodern" politics, then, do not seem unprecedented, dramatically new, or original. If there is truth in these descriptions, as I think there is, their power and importance are not diminished if, instead of characterizing them as radical and unprecedented developments, we see them instead as a fuller expression of tendencies already present.

The second point is that postmodern forms and practices have not taken over as totally as is depicted in some works on postmodernity. Cogent as these descriptions are, the most cursory examination of contemporary events suggests that they describe only part of what is happening politically in American society. Indifference to public matters may be endemic, but it is far from total and uniform. Citizens vary widely in their levels of attention to and concern about issues. Earlier forms of political practices, traditional social movement forms and strategies, for example, persist. Contemporary society should be seen as complex and heterogeneous, as having a culture that combines traditional, modern, and postmodern elements. Although some theorists of postmodernity totalize what they describe, others agree that postmodernity is only one facet of a complex and multilayered social formation and are groping toward a careful and detailed theorization of uneven and combined cultural development. See, for example, Jameson, 1991:3-4; Lash, 1990:13.

39. The image/word distinction is discussed in detail by Lash, 1990:172-198.

40. It is, of course, hard to have read Durkheim, Saussure, and the French structuralists without wondering how something that was *always* interpenetrated with and constitutive of

social reality—that is, the organization of collective representations—can now have expanded to where it wasn't before.

41. See, for example, Jameson's discussion of how postmodernity transforms people's relationship to their own history (1984).

42. Jameson, 1984, offers an especially thorough discussion of postmodern subjectivity.

43. Kaplan labels these "utopian" versus "co-opted" postmodernisms (1988:4). Lash calls them "oppositional" versus "mainstream" postmodernisms (1990:37).

44. Erikson, Luttbeg, and Tedin, 1980; Jones, 1990; Neumann, 1986; Oreskes, 1990c.

45. The second definition of *icon* in *Webster's Ninth New Collegiate Dictionary* is as follows: "a conventional religious image typically painted on a small wooden panel and used in the devotions of Eastern Christians." A subject search on the on-line library catalog at the University of California, Berkeley, brings up titles concerned exclusively with the religious connotation of *icon*: Balkan icons, Carpathian, Albanian, Byzantine, Greek, Byelorussian, Bulgarian icons, and so on.

46. See Sturrock's review of the different types of signs (1986:84-85).

47. Politics has always relied heavily on slogans to carry emotionally loaded, condensed meanings in the longer signifying chains of political discourse. I am talking here about a *tendency*, not an absolute, a tendency in which the discursive rhetoric fades out, is no longer performed, leaving the visual image and the sound bite to carry all the freight of the political message by themselves. Postmodern theorists' depictions, here, converge with the laments of the political pundits who say that the mediafication of politics has altered every aspect of the political process, mutating it in form and degrading it in quality: electoral campaigns run along the lines of commercial product advertising; candidates' positions on "the issues" trivialized, "communicated" with a few carefully chosen images and sound bites; lawmakers continuously preoccupied with the politics of image management (Oreskes, 1990a, 1990b).

48. Dunlap, 1986; Freudenburg and Baxter, 1983; Iyengar and Kinder, 1987:31; Mazur, 1981, 1984b; Rankin, Nealey, and Melber, 1984; Schoenfeld et al., 1979.

49. Starr, 1980; Wildavsky, 1979.

50. Downs, 1972; Dunlap, 1986; Iyengar and Kinder, 1987; Schoenfeld et al., 1979.

51. One possible way to think about this twofold nature of viewers' reactions is in terms of poststructuralist conceptions of "subject position." What subject positions are made by media coverage of contamination events? On one hand, it allows people to identify with victims, feel for them, and feel what it would be like to be them; on the other hand, the whole format of watching bad news happen elsewhere while one is sitting safe, in fact untouched, in the unchanging environment of one's home creates subject positions that make a distinction between private and public, safe here and unsafe some other place. See, for example, the interview with Kaye Kiker in chapter 8 of this volume: "I was more or less not interested. It just seemed like a lot of bad news. I didn't pay a lot of attention to the news. The newspapers—I really didn't read 'em. I was just too busy. Just always figured that, figured it was the same news, it just happened to different people."

52. This took the form of the Superfund law, the Comprehensive Environmental Response, Compensation and Liability Act of 1980, P.L. 96-510. Discussion in this section is based mostly on Senate, 1983. Unless otherwise noted, the citations in this section refer to those volumes.

53. At the time, it was estimated that Love Canal would cost $30 to $125 million (Senate, 1983, vol. 1:318). Ten years later, the cost stood at $275 million—just for studying the site, covering it, and buying the homes around it (Shabecoff, 1990b).

54. "The CMA and its member companies strongly support new legislation to solve the problem of abandoned hazardous waste sites" (Dr. Louis Fernandez, vice chairman of Monsanto, on behalf of CMA, Senate, 1983, vol. 2:237). In Congress, only the right fringe dared categorically oppose a basic cleanup program (Senate, 1983, vol. 1:724; vol. 2:245, 328).

55. Whatever the size of the fund, billions more would be needed to pay for the cleanup of all Superfund sites. Unless outlays from the fund could be recovered from responsible firms, the fund would be depleted long before a significant number of cleanups could be accomplished. Traditional liability doctrine would make recovery difficult. Firms could be found factually responsible for pollution but still not be held liable unless victims could show the firms had been negligent. Furthermore, it would often be the case that several firms had contributed wastes to the same problem site. Traditional liability doctrine placed upon the plaintiff, the EPA or the private victim, the burden of apportioning the damage among defendants. Lawmakers concerned about this problem found two innovations in liability doctrine appealing. "Strict" liability establishes liability without the need to show negligence; a showing of factual responsibility suffices. The doctrine of "joint and several" liability addresses cases in which the injury is indivisible or, practically, cannot be apportioned among several wrongdoers. Imposing joint and several liability on hazardous waste activities would make recovery easier because the government would no longer bear the burden of establishing how much each firm contributed to contaminating the site in question; *any* one party, even if only partially responsible, could be held liable for the whole site.

56. "Existing state tort laws present a convoluted maze of requirements under which a victim is confronted with a complex of often unreasonable requirements with regard to theories of causation, limited resources, statutes of limitation and other roadblocks that make it extremely difficult for a victim to be compensated for damages" (Albert Gore, in Senate, 1983, vol. 2:94-95).

Liberals in Congress proposed a number of changes that would improve victims' likelihood of achieving success. First, the liability provisions would specify that polluters would be liable not only to the fund, but also *to victims* (see, for example, ibid.:40-41). Such a provision would create what was known as a "federal cause of action," a federal toxic tort, that would allow injured parties to pursue a remedy in the federal courts, while imposition of strict, joint and several liability would make it easier to do so successfully. Second, liberals advocated what was referred to as "medical causation" language that would statutorily change the rules of evidence to allow introduction of, for example, statistical/epidemiological studies and tests on laboratory animals and microorganisms as evidence that proves causation.

57. Conservatives argued that the feedstock tax "unfairly places the burden on today's companies, shareholders, or customers for practices, failures or shortcomings of yesterday's industrial producers" (Senate, 1983, vol. 2:100). The chemical industry felt the feedstock proposal was "simply unfair" (ibid.:237).

Moderates joined conservatives in opposing innovations in liability doctrine: "S. 1480 . . . simply throw[s] out negligence and nuisance as legal concepts relating to the ongoing activities of America's entire industrial base. . . . creating liability that will adversely affect the decisionmaking process of American business. The liability provisions of this bill alter . . . the potential exposure of the company's assets to loss. . . . it prevents the private sector decisionmaking process from being able to assess its liability risks. . . . adverse consequences for the national economy. . . . Small companies may be forced to close. . . . Middle-sized companies will not likely risk their assets for new ventures. . . . Large corporations, faced with the alternatives of building or expanding facilities in this country or elsewhere, may choose to spend their capital funds in countries with less possibilities of risk" (additional views of Senators Domenici, Bentsen, Baker, Senate, 1983, vol. 1:427-429; for more extreme views, see David Stockman's speech, vol. 2:358).

Tort reform proposals, too, were opposed by conservatives, both as part of the larger opposition to changing the liability standards and in their own right. Domenici, Baker, and Bentsen phrased their concerns thus: "The bill creates new judicial standards for presentation of evidence and pursuing a cause of action, the combined effects of which are unknown and untested. S. 1480 . . . allows individual claimants to enter court more easily and to proceed with a suit despite a paucity of evidence. . . . the bill significantly reduces the need for a plaintiff to establish a causal link between a hazardous substance and an alleged injury. . . . constitutes an intrusion into judicial processes that have been formulated over hundreds of years of common law evolution and procedural developments upon which industries and businesses have relied in assessing their exposure to liability" (vol. 1:426).

58. "We recognize that there is a gap in existing law for dealing with . . . orphan hazardous waste disposal sites. We would support legislation that would close this gap" (dissenting views of Reps. Broyhill, Devine, Collins, Loeffler, Stockman, in Senate, 1983, vol. 2:98).

"We do support legislation to deal with this problem of cleaning up abandoned and inactive hazardous waste sites. . . . and we do not like to hear, because we do question certain sections, that we are opposed to the bill or trying to block the legislation" (Rep. Broyhill on the House floor, in ibid.:241).

"I have stated publicly on many occasions that I was fully prepared to work for the passage of an abandoned dump site act to deal with the issue of 'orphan dumps.' . . . These are potential killers in every sense of the word" (minority views of Sen. Simpson, in vol. 1:423-424).

59. Just how fragile and uncertain things were is shown by the letter sent by Senators Stafford and Randolph to the House of Representatives. The senators begged the House leadership to adopt the Senate version of the bill, thus avoiding a House-Senate conference: "Our bill was brought to the Floor not because a large number of Senators supported it, but because a large number agreed not to oppose it. The bill represents an extremely delicate balancing of interests. . . . it made its way through the Senate only because there was unanimous consent that there would be no changes to the compromise . . . and because virtually every Member agreed on the need for a bill. . . . Had we changed a comma or a period, the bill would have failed. . . . That the bill passed [the Senate] at all is a minor wonder. Only the frailest, moment-to-moment coalition enabled it to be brought to the Senate floor and considered. Indeed, within a matter of hours that fragile coalition began to disintegrate to the point that, in our judgement, it would now be impossible to pass the bill again, even unchanged. . . . With the evaporation of the balance of interests . . . amendment to the bill will kill it if it is returned to the Senate. . . . it was the *only* bill we could pass at the time and we do not believe it can be passed again" (in Senate, 1983, vol. 1:774-775).

4. The Toxics Movement: From NIMBYism to
Radical Environmental Populism

1. EPA, 1979. Early contamination protests in Durham, Connecticut (1970), Shenandoah Stables, Missouri (1971), and Woburn, Massachusetts (1972), are described in Brown and Mikkelsen, 1990; Epstein et al., 1982.

2. This in spite of the fact that, by their own admission, problems with their case-finding technique led them to underestimate the extent of public opposition (Duberg, Frankel, and Niemczewski, 1983-84:84).

3. EPA, 1974; House of Representatives, 1976.

4. Brown, 1979; Epstein et al., 1982; Freudenberg, 1984; Nader, Brownstein, and Richards, 1981.

5. That one case was in El Dorado, Arkansas (EPA, 1979:283).

6. The difference in the success of contamination organizing before and after Love Canal can be seen in cases such as those in Memphis (Thomas, 1981) and Toone, Tennessee (Editors, *Southern Exposure*, 1981b).

7. The best source of case histories complete enough to judge presence or absence of contact with other communities is EPA, 1979. Some examples of early networking include the following: At El Dorado, Arkansas, opponents of Ensco's proposal to burn PCBs brought in people who had fought an Ensco facility in Minnesota (ibid.:286). Opponents of an SCA facility at Model City, New York, developed ties with people in Love Canal and brought in the

Illinois assistant attorney general who had prosecuted the SCA facility at Wilsonville, Illinois (ibid.:268-269). Opponents at a proposed SCA facility in Bordentown, New Jersey, brought in speakers from Wilsonville, Love Canal, and Model City (ibid.:164).

8. "For the vast majority of groups in the Movement, the local fight is everything" (Lois Gibbs, in Zeff, Love, and Stults, 1989:39). "People that see themselves in their fight often don't want to join a larger group [i.e., state coalitions] where there are more meetings and fight because it becomes overwhelming" (Sally Teets, interview). Sue Greer talked of being "very harsh with some groups" because they do not want to keep fighting, helping others, but she noted, "it does happen a lot [that they quit] because people get tired of working on the issue" (interview).

9. Here is a sample of the responses from interviewees for this volume when asked if they kept records of the numbers of groups they work with. Will Collette, CCHW's national organizing director, said, "Never in our ten-year history have we ever had enough staff to meet demand." Sally Teets, CCHW Midwest organizer, had the following exchange with interviewer Hal Aronson:

Q: How many groups do you work with?

A: I wouldn't have any way of even estimating. . . . My region is twelve states that range from Ohio to Montana. . . . If I went to everything I was invited to I would never be home with my family.

Q: It seems like an enormous work load.

A: It is! . . . there are only six of us in the field and we are all in single-person offices. We don't have any support staff.

Q: You're it, you're CCHW.

A: I'm it. I am CCHW in the West. There is nobody here but me. The phone rings here in my home and I work out of my basement. I have no secretary, no go-fers, no filers.

Q: This sounds like a huge job.

A: Forty hours a week plus being on the road. I had been, for months, in the summer, on the road for fifteen out of thirty days a month.

Sue Greer, from PAHLS, the Indiana coalition, answered as follows:

Q: Do you know how many local groups PAHLS has worked with or helped?

A: Oh my God. . . . I would say, I don't know, I'm sure it's fifty or more.

Q: Fifty over the last nine years or currently?

A: That is so hard for me to say . . . I just have no way of knowing the number and I'm going to tell you why: We don't—I was criticized by somebody for that one time. He said, "You should make a record of that." And I said, "Why should I write it down?"—I mean that's not my goal to have this giant list of kudos, and so I don't look at it that way. I don't think anybody else in our group does.

Outsider researchers could analyze these groups' raw phone contact sheets and the like. The records were not made available to me for independent analysis.

10. CCHW, 1984, 1985, 1986a, 1987a, 1988.

11. CCHW, 1988. In 1989, CCHW's quarterly publication, *Action Bulletin* (nos. 21-24) carried reports of local fights in forty-four states and the District of Columbia.

12. See Sally Teets, note 9, above. Marty Chestnutt, CCHW South organizer, when asked, "How many groups are you working with?" answered, "Ah, gee. It varies. Probably right now about thirty . . . all grass-roots groups. . . . Of course, now, tomorrow it may be forty. . . . Today I heard of four more landfill groups that I'm going to be in touch with before tomorrow and so that will be four more groups that are added, but then I may lose two or three the next day or two as they win their fights. So it's ongoing. It's very flexible. And I guess if you started talk-

ing about all the groups that I keep in touch with, you're probably talking about four or five hundred."

Organizers for other groups also report large work loads, confirming the impression that there are many, many local struggles taking place. Sue Greer of PAHLS notes in her interview, "Our phone rings. It keeps three people in here every day just to answer the phone. We get no less than fifty or sixty calls a day." Lew Dunn, an organizer in California for Greenpeace, said, "In 1987, I think it is when I went on my first one. In Amador County. . . . From there I went to Maricopa, to Avenel, Kettleman, Benicia, Richmond, Martinez, Tehama County, Redding, San Rafael, Newberry Springs, Ludlow, Alpaugh, McFarland, Kern County, Shafter, Susanville, Yerrington (Nevada), Lone Pine, Bishop . . . [Q: "These are in 1987?"] These are last year [1989]. I've done that with Greenpeace in the last year. That's a lot. I put 70,000 miles on my car" (interview). John Thompson of the Central States Resource Center (CSRC) said in an interview that he has worked with people fighting about 130 sites in six years. These are mostly solid waste landfills, some hazardous waste landfills, some Superfund sites.

13. Interviews with Gary Cohen, National Toxics Campaign, and Margie Kelley, Greenpeace.

14. The quotes presented here are just a selection. The full set of citations in the literature that refer to local action as the biggest problem also include statements by officials from major generator corporations (Farkas, 1980), other statements by waste industry officials (Buckingham, Buchanan, and LaGrega, 1986; Pirages, 1987), waste industry consultants (McCoy and Associates, *Hazardous Waste Consultant*, all issues); a multisector siting problem-solving workshop brought together by the Conservation Foundation (Hazardous Waste Dialogue Group, 1983), the EPA and EPA officials (EPA, 1980, 1982a, 1982b; *Environment Reporter*, 1979, 9:2114–2115, 2295), state officials (Canter, 1981, 1982; Dodd, 1986; Galida, 1984; McGuire, 1986; Vince, 1982; and my own telephone survey of twenty-two state waste agencies), and almost every policy scientist who studies hazardous waste policy (Bacow and Milkey, 1987; Bingham, 1984; Elliott, 1984; Morrell, 1987; Morrell and Magorian, 1982; O'Hare, 1977; O'Hare, Bacow, and Sanderson, 1983; Piasecki and Davis, 1987; Popper, 1987a, 1987b; Sandman, 1987; Tarlock, 1984).

15. "When I heard about Stringfellow, I went down to Penny Newman. What happens is that you are battling at home and you get to the point where you say, 'Jesus, is this happening everywhere? Why am I having such a hard time winning here?' . . . I went to Stringfellow and talked to her. I went to IT Benicia and talked to Marilyn O'Rourke's husband. I went to Kettleman. And I found out that we all had the same thing in common" (Lew Dunn, Casmalia, California, resident, now community organizer for Greenpeace, interview).

16. "Once they win, then they definitely will want to share the knowledge they've learned with other people. . . . [There is] a tremendous amount of wanting to share" (Marty Chestnutt, CCHW South, interview).

17. Sally Teets: "I take [a] grass-roots group . . . and teach them how to put on a lot of political pressure. . . . I come in and do motivational speeches for the community, tell them how I won our local grass-roots fight . . . remotivate them. Put on training sessions to teach organizing, fund-raising, dealing with politicians, dealing with media" (interview).

Marty Chestnutt: "When a group calls and they say 'help!' . . . whatever their problem, they call me and they want to know what to do next and then I give them information and I walk them through how to organize and I'm always here if they have a problem, if they run up against a blank wall and they don't know what to do next. . . . and that's basically what I do. I answer questions and I calm fears and I give 'em pep talks when necessary" (interview).

18. For a description of what happens at a Leadership Development Conference, see Zeff and Greer, 1988.

19. An example of several local groups, initially formed independently of each other, finding each other and forging a regional coalition is TEACH, Tennesseeans Against Chemical Hazards (Editors, *Southern Exposure*, 1981b:46). Some examples of a single strong group

becoming a regional organizing center include Independent Citizens Associated for Reclaiming the Environment (I-CARE), Ohio (Kraft and Kraut, 1988); Citizens Alliance for a Safe Environment (CASE), New England and the Middle Atlantic states (Bush, 1981:534); Georgia 2000 (Thomas and Brooks, 1981:41); and PAHLS, Wheeler, Indiana. Concerning "top-down" coalition formation: "The national groups, including CCHW, have instigated 'external hubs.' We have done it by strategic choices about where we would have LDCs. 'Gee whiz, we are getting mountains of requests for individual group help out of this particular state. Let's hold a Leadership Conference so we can try to do it on a wholesale basis instead of retail and let's keep our fingers crossed and hope that something good and cohesive will come out of it' " (Will Collette, interview).

I was able to identify coalitions in twenty-six states (interviews with Will Collette and Gary Cohen; and CCHW, 1986b). Will Collette noted, "There isn't a state in the Union where there isn't some thought or consideration being given to moving in this direction because it is a natural evolutionary process."

20. CCHW, 1986b; Lewis and Kaltofen, 1989; National Campaign Against Toxic Hazards, 1988; NTC, 1989, n.d.

21. For critiques of traditional environmentalism's relationship to issues of race and inequality, see Anthony, 1990; Bullard and Wright, 1990; Lewis, 1990.

22. Bryant and Mohai, 1992.

23. Commission for Racial Justice, 1987; Bullard, 1990; GAO, 1983b; various issues of *Race, Poverty and the Environment*, published by the Earth Island Institute.

24. See chapter 5.

25. NTC, *Toxic Times*, 2(2)10; *RACHEL*, 48; Smothers, 1991.

26. Albany State College Toxics Communications and Assistance Project, 1989; Thomas and Brooks, 1981; Wurth-Hough, 1982. See also interviews with Will Collette and Marty Chestnutt; *Race, Poverty and the Environment*, 1990, 1(3):11.

27. CCHW, *Everyone's Backyard*, 1989, 7(1):2; Will Collette, interview; NTC, *Toxic Times*, 2(2):10; Smothers, 1991. In 1990, another Native American organization, Citizens Against Ruining Our Environment (CARE), hosted a meeting of twenty-six tribes to discuss toxic dumping on their lands. *Race, Poverty and the Environment*, 1990, 1(3):9.

28. *Race, Poverty and the Environment*, 1990, 1(2):12, 18.

29. Luke Cole, interview. For the latest developments in the burgeoning area of "environmental racism," readers should consult recent issues of *Race, Poverty and the Environment*.

30. Greenpeace, 1988; Sarokin, 1987; also, Margie Kelley, Lew Dunn, and Will Collette, interviews.

31. The Sierra Club took on the toxic waste issue fairly early—Epstein, Brown, and Pope's influential book on hazardous waste was published in 1982 by Sierra Club Books—and continues to give the issue some prominence. See, for example, Sierra Club, 1987. The Environmental Defense Fund (EDF) and National Resources Defense Council (NRDC) give the issue considerable attention. See EDF, n.d., 1986; interview with Lois Epstein of EDF; interview with Don Strait of NRDC. Others, such as Audubon, National Wildlife Federation, and the Conservation Foundation, devote fewer resources to the toxic waste problem, but do mention it in their literature, take part in the process when legislation is up for reappropriation, and support the idea of waste reduction. National Wildlife Federation, 1986; interviews with Gerry Poje, Wildlife Federation; Ann Strickland, Audubon; Lane Krahl, Conservation Foundation. I should note here that some other, more broadly defined, liberal political organizations, such as the League of Women Voters and USPIRG, also have been quite involved in toxic waste policy. Interviews with Lloyd Leonard, League of Women Voters; Rick Hind, PIRG.

32. From various interviews, it seems that there is some distrust and tension between the traditional environmental organizations and the newer toxics movement organizations. What strikes the outsider, however, is that in spite of the mutual distrust and disapproval, the two types of environmentalisms have, in practice, worked out a very effective division of labor.

Traditional environmental organizations would have less political clout were it not for the grass-roots movements that tell politicians that mass sentiment for these issues is real. Conversely, the policy gains that Washington-based environmental organizations secure contain access-to-information provisions and public-participation provisions that help local organizing. Such tense, subjectively antagonistic but mutually beneficial relationships are not uncommon within what McAdam, McCarthy, and Zald refer to as "movement industries" (1988: 718-719).

33. CCHW, 1986b:20, 29-31.

34. CCHW, 1988, 1989:57-58.

35. Sometimes these charges themselves border on the hysterical; for example: "The NIMBY syndrome is a public health problem of the first order. It is a recurring mental illness that continues to infect the public. Organizations that intensify this illness are like the viruses and bacteria which have, over the centuries, caused epidemics such as the plague, typhoid fever, and polio. . . . It is time solid waste management professionals stopped wringing their hands and started a campaign to wipe out this disease" (*Waste Age*, March 1988:197; quoted in *RACHEL*, 109).

36. Lois Gibbs, in CCHW, 1989; CCHW, *Everyone's Backyard*, 1990, 8(1):2; CCHW, 1989:48; Sue Greer, in *PAHLS* 6(3); National Campaign Against Toxic Hazards, 1988:II-14; NTC, n.d.:1.

37. *PAHLS*, 1989, 7(1); 1989, 7(3); CCHW, *Everyone's Backyard*, 1990, 8(3):22; Kaye Kiker, Sue Greer, interviews.

38. *PAHLS*, 1988, 6(1). The Silicon Valley Toxics Coalition sells poster-sized prints of Chief Seattle's letter.

39. CCHW, 1989:57-58; NTC, n.d.:5-6. See also NTC, 1989; Sarokin, 1987. For other voices in the movement, at the state and regional coalition level, see *PAHLS*, 1988, 6(3); Silicon Valley Toxics Coalition's *Silicon Valley Toxics News* (any issue); Texans United's *Texas Report*, 1989. *RACHEL's Hazardous Waste News* repeatedly emphasizes the centrality of source reduction; see, for example, *RACHEL*, 5, 16, 33, 81, 54, 111, 154, 155.

40. In chapter 8 I will describe how the center works to radicalize the base.

41. In this discussion of the social movements literature, I have found McAdam et al.'s (1988) survey most helpful.

42. By now, we have a substantial body of research on the experience of contamination. Edelstein, 1988; Fowlkes and Miller, 1982; Gibbs, 1982; Kroll-Smith and Couch, 1991; Levine, 1982. I use Kroll-Smith and Couch's work extensively in this discussion.

43. Kroll-Smith and Couch refer to this as the development of "threat beliefs" (1991:30-33).

44. We might add another, historically more specific, source of contemporary moral economy, as well. Shortly after World War II, management and organized labor in the United States struck a grand bargain in which the unions would give up conflict over conditions of work in exchange for promise of routine and more favorable agreements over compensation issues. Implicitly, the bargain meant a deal in which workers would be willing to have little or no power in the workplace over such things as conditions of work, technological innovations, and the process of production in exchange for more money—in effect, an exchange of greater alienation at work for consumer pleasure and the right to be left alone to enjoy one's "private" life at home. This is, of course, violated in the extreme when one finds out that the toxic workplace is now seeping into one's basement and into one's children's bodies. For another type of discussion of the role of "the activation of moral norms against harming innocent people" as a basis for support for environmental protection, see Stern, Dietz, and Black, 1986.

45. "The victims of off-site releases of toxic contamination are almost uniformly disappointed, frustrated and angered by their perceptions that responsible parties will not admit their culpability, nor act swiftly and efficiently to abate the hazard, clean up the site and compensate the victims" (Kroll-Smith and Couch, 1991:9).

46. Recall Will Collette's comment, quoted above, about the relationship between media coverage and people's new willingness to believe they, too, suffer from contamination: "The consequence of Love Canal and Love Canal getting a tremendous amount of media play was that all of a sudden out of the woodwork hundreds of citizen groups started to form, very spontaneously, because all of a sudden Love Canal sort of gave rise to what they had been thinking and feeling" (interview).

47. EPA, 1979, 1980; GAO, 1978a; U.S. Council on Environmental Quality, 1980. My discussion of siting opposition is based on the numerous analyses of the causes of opposition in the waste policy literature. Bacow and Milkey, 1987; Cerrell Associates and Powell, 1984; EPA, 1979, 1980; Keystone Center, 1980, 1984; Morrell, 1987; Morell and Magorian, 1982; O'Hare, 1977; O'Hare, Bacow, and Sanderson, 1983; Popper, 1987a, 1987b; Sandman, 1987; Tarlock, 1984. Of these, EPA, 1979, is the most convincing, because Centaur Associates, EPA's consultants, based their analysis on direct, detailed examination of a large number of cases.

48. Some examples: "People in Spencerville are 'tired of being crapped on' by Waste Management" (*RACHEL*, 51). The Emelle, Alabama, hazardous waste landfill was dubbed "America's pay toilet" in *Greenpeace*, 1987, 12(3):13. "One woman who lives in the farming community put it very simply, she said that as a dairy farmer, she sells her milk and beef to the market. She doesn't expect the consumer to take care of the cow manure, why is it then that we must sit and discuss what to do with industry's manure? We no more want industry's waste in our backyard than industry would like our manure in their front yards" (Lois Gibbs, testifying in House of Representatives, 1982a:18).

49. Kroll-Smith and Couch, 1991:32. The authors confirm, however, that the impulse to flee was never really an effective or workable response (ibid.:17). For most people who are dealing with potential contamination, their homes are their only form of accumulated savings. Now they can't sell those homes. Remember the reporter asking Mary Heeney, a Love Canal resident, "Have you thought about trying to sell [your home]?" She laughed bitterly and answered, "They won't put for sale signs on any of these homes and who'd be crazy enough to buy one? I can't imagine why anybody would." Most contamination victims are unlikely to have the income to allow them to keep up mortgage payments on an empty house while they pay rent elsewhere. In addition, people find it difficult to face the prospect of those other, nonmonetary, costs that they incur if they leave, such as the disruption of the sustaining routines of everyday life and the loss of friends.

50. A recent study suggests that the evidence on community costs and benefits is considerably more equivocal than previously thought (Smith, Lynn, Andrews, Olin, and Maurer, 1985), but the salient fact here is people's *perception*. The literature on compensation as a way of artificially providing some kind of benefit and in that way increasing the likelihood of community acceptance—see chapter 5—tends to support the notion that, to citizens, these facilities seem to have no intrinsic benefits.

51. For Reagan deregulation and Sewergate, see chapter 6. For implementation of Superfund, HSWA, and SARA, see chapter 7. For comments on the adverse impact on public trust of the Reagan administration's treatment of RCRA and Superfund, see testimony by Joel Hirschhorn of the Office of Technology Assessment, in Senate, 1982:9, and by Randy Mott of the Hazardous Waste Treatment Council, ibid.:39.

52. In contrast to cases of contamination, where organizing is hindered by the inherent ambiguity, uncertainty, and invisibility, siting opposition is helped by the fact that hazardous waste facilities are both visible and concentrated. Unlike some other important environmental threats that inherently lack visibility and are diffuse—ozone depletion, for example—here the threat is embodied in a *facility*. Facilities are a presence. They are major construction projects. When operating, they generate constant sensory impacts: odors, truck traffic, drums with hazard warnings stenciled on them, employees who look like they are suited up for space travel.

53. EPA, 1979, documents that existing organizations and local leaders played a key role in some cases of siting opposition. On the other hand, consider Love Canal, or Sue Greer and PAHLS. There is plenty of evidence that time and again local toxic waste protest organizations are built in the social movement analogue of virgin territory.

54. The phrase is from Mike Davis, cited by Callinicos, 1989:164.

55. Gibbs, 1982; Levine, 1982.

5. Could Opposition Be Neutralized? Discourses and Policies of Disempowerment

1. P.L. 98-616, 98 Stat. 3221. Passage and provisions of this act are discussed in chapter 6.

2. In 1979 and 1980, the EPA estimated that 50 to 125 new sites would be required. See *Environment Reporter*, 1980, 11:272; EPA, 1980:3; GAO, 1978a:ii. Later developments suggest that alarms about capacity crisis may have been exaggerated. Although hardly any new capacity came on line during the 1980s, an *overall* capacity problem, on a national scale, failed to develop (Alm, 1984:1; Legislative Commission on Toxic Substances and Hazardous Waste, 1987:41). In spite of that, specific firms and industries could well have been troubled by regional shortages or the paucity of disposal service at a good price.

3. For concurring comments about public trust, see Cerrell Associates and Powell, 1984:7; EPA, 1980:3.

4. See chapter 6.

5. Recall the quotes provided in chapter 4.

6. GAO, 1978a:11-12.

7. Wes-Con successfully used abandoned Titan missile silos in Idaho to dispose of toxic wastes (EPA, 1979:144).

8. See references in chapter 4 notes 22 and 23.

9. Anderson and Greenberg, 1984:183; GAO, 1978b:1.

10. Cerrell Associates and Powell, 1984:app. C.

11. Editors, *Southern Exposure*, 1981a, 1981b; EPA, 1979; Freudenberg, 1984; Geiser and Waneck, 1983; Thomas and Brooks, 1981.

12. Canter, 1981:814; National Conference of State Legislatures, 1982.

13. Morrell and Magorian, 1982.

14. Critiques and suggestions about how to make the preemption idea more workable can be found in Andrews, 1988; Bacow and Milkey, 1987; Bingham, 1984; Keystone Center, 1980, 1984; McGuire, 1986; Morrell, 1987; Morrell and Magorian, 1982; O'Hare et al., 1983; Tarlock, 1984.

15. O'Hare, 1977:409, 419.

16. For discussions of the varieties of compensation, see O'Hare et al., 1983:72-73; EPA, 1982a:5-16.

17. Polanyi, 1944:103-129. Polanyi quotes from Daniel Defoe's 1704 pamphlet, *Giving Alms No Charity and Employing the Poor a Grievance to the Nation*: " 'Giving alms no charity' — for in taking away the edge of hunger one hindered production and merely created famine; 'employing the poor, a grievance to the nation' — for by creating public employment one merely increased the glut of the goods on the market and hastened the ruin of private traders" (109).

The early political economists agreed that the poor were needed to do society's hard and dirty work in peace and to be its cannon fodder in war: "For if one had a hundred thousand acres of land and as many pounds in money, and as many cattle, without a labourer, what would the rich man be, but a labourer? And as the labourers make men rich, so the more labourers there will be, the more rich men . . . the labour of the poor being the mines of the rich" (John Bellers, 1696, quoted in Marx, 1867/1967:614).

"It would be easier . . . to live without money than without poor; for who would do the work? . . . besides, they are the never-failing nursery of fleets and armies, without them there could be no enjoyment, and no product of any country could be valuable" (Bernard Mandeville, *Fable of the Bees*, 1728, quoted in ibid.:614-615).

"For what is it but distress and poverty which can prevail upon the lower classes of the people to encounter all the horrors which await them on the tempestuous ocean or on the field of battle?" (William Townsend, quoted in Polanyi, 1944:118).

Michael O'Hare runs with an impressive crowd.

18. Bacow and Milkey, 1987; Legislative Commission on Toxic Substances and Hazardous Wastes, 1987; Portney, 1988.

19. Bingham, 1984; Keystone Center, 1980, 1984; Tarlock, 1984.

20. Ethridge, 1987; Sewell and O'Riordan, 1976.

21. For similar comments, see Bingham, 1984:18; Keystone Center, 1980:13; EPA, 1979:9.

22. Supporters of expanded participation as the key to success in siting included the major corporatist dialogue efforts that brought capital, government, and environmental organizations together to develop a workable siting model (Keystone Center, 1980, 1984; Hazardous Waste Dialogue Group, 1983); state officials and consultants (Canter, 1981; Cerrell Associates and Powell, 1984; Dodd, 1986; McGuire, 1986); and most of the important policy scientists working in this field (Andrews, 1988; Kraft and Kraut, 1988; Matheny and Williams, 1988; Morrell, 1987; Morrell and Magorian, 1982; Piasecki and Davis, 1987; Sandman, 1987; Susskind, 1985; Tarlock, 1984). The discussion below is based on these, as well as on Bingham, 1984; Clapham, 1985; Craig and Lash, 1984; EPA, 1979; Farkas, 1980; Forcade, 1984; Robbins, 1982; and Windsor, 1987.

23. For similar advice, see Windsor, 1981:523.

24. This important distinction is discussed by Habermas, 1984.

25. The EPA published a series of brochures on siting strategies, issued a Model Siting Statute that states could use as a model, and wrote a strong letter to state governors urging them to action. See Canter, 1982; EPA, 1980, 1982a, 1982b.

26. Andrews, 1988; Bowman and Lester, 1985; Hadden, Veillette, and Brandt, 1983; Legislative Commission on Toxic Substances and Hazardous Wastes, 1987; National Conference of State Legislatures, 1982; Piasecki and Davis, 1987; Ryan, 1984; Tarlock, 1984. Siting statutes vary from the extremely passive to the strongly active. The passive approach leaves it to the private sector to initiate proposals for new facilities, and local governments retain significant power to approve or deny proposals. In the active approach, the state develops a statewide management plan, initiates the siting process by identifying locations suitable for a facility and actively soliciting private sector participation, and preempts all local authority. The active approach is described in Hadden et al., 1983:199-206; Legislative Commission on Toxic Substances and Hazardous Wastes, 1987:10; Piasecki and Davis, 1987:280-281. Some states went so far as to mandate that a facility *must* be sited; a few were willing to own and operate a state facility. Overall, surveys indicate that strong initiative is "a minority approach to siting" (Tarlock, 1984:121) and that most states opted for the more passive end of the spectrum. "Generally, states appear to prefer the role of guardian to the siting process while allowing the private sector to set the pace and initiate specific siting proposals" (Legislative Commission on Toxic Substances and Hazardous Wastes, 1987:14). Similar conclusions are reached by Bowman and Lester, 1985. My discussion of the kinds of provisions found in state statutes is a summary based on these reviews.

27. On moratoriums, see Legislative Commission on Toxic Substances and Hazardous Wastes, 1987:29. Blanket bans of out-of-state wastes were ruled unconstitutional by the federal courts (*City of Philadelphia v. New Jersey*, 1978; *Hardage v. Atkins*, 1980). However, in 1989, Alabama and South Carolina tried another tack by banning the import of wastes from states that have no disposal or treatment capacities of their own (Associated Press, 1989). By

the mid-1980s, federal lawmakers were convinced that the states were not standing tall in the face of public sentiment and that some measures had to be taken to get state governments to get tougher on siting. They came up with one threat in the 1986 Superfund reauthorization. In that statute, Congress required that any state wishing to continue to get Superfund money to clean up old sites within its boundaries produce a twenty-year capacity assurance plan by 1989 (Gold, 1989; Smothers, 1989).

28. A 1984 Massachusetts survey identified thirty-two attempts to site new commercial off-site facilities. Of those, eight had been approved (though not necessarily built), eighteen proposals had failed, and six were still pending (Ryan, 1984:11). A 1987 New York survey identified eighty-one siting attempts nationally. Of those, thirty-six were still pending, thirty-one had failed or were withdrawn, and fourteen were permitted; of that fourteen, six had actually been built and become operational (Legislative Commission on Toxic Substances and Hazardous Wastes, 1987:28). A 1986 EPA survey found somewhat higher numbers: of 179 attempts to site, 25 percent were failures, 22 percent were successes, and 53 percent were pending (cited by Andrews, 1988:120; McCoy and Associates, 1985, 1986, 1987, 1988, 1989, 1990).

29. As of 1990, EPA could not provide a reliable and complete list of facilities that had been built and permitted since 1976 (Emilie Schmeidler, private communication).

30. EPA, 1980:3; *Environment Reporter*, 11:272-273, 1980.

6. Hazardous Waste Regulation Progresses against the Conservative Tide

1. Such corporate behavior in the case of RCRA was described in chapter 2. See also Epstein et al., 1982:181-256. For similar behavior in the case of worker safety and health regulation, see Szasz, 1982, chs. 3, 4; 1984.

2. Stone, 1981.

3. MacAvoy, 1979; Weidenbaum, 1979. Liberal economists took a somewhat more moderate position, but they, too, agreed that too much of the traditional type of regulation was bad for the economy and that fundamental reforms were necessary. See Lave, 1981. Economists' views were quickly embraced by business leaders. See Conference Board, 1980. It should be noted, however, that the causal connection between "too much" regulation and the problems that plagued the U.S. economy was far from clear. See, for example, the exchange among three economists in Martin and Schwartz, 1977:4, 24, 27. MacAvoy, a leading advocate of deregulation, conceded, in 1979, a year before deregulation was officially declared to be one of the main pillars of economic recovery, that the impact of pollution controls on GNP growth has "probably not been substantial" (1979:92). For a critical examination of the purported connection between economic harm and social regulation, see Szasz, 1986c:31-34.

4. U.S. Bureau of the Census, 1980:407, 438, 477.

5. Converse et al., 1980:235-236, 248-249; Gallup, 1972, 1978, 1979, 1980, 1981.

6. For examples of speeches by chairmen of major corporations, see Editors, *Business Today*, 1977, 1978; Melloan and Melloan, 1978:237-238.

7. CBS/New York Times polls, done in 1978 and 1981; Harris Poll, 1976, 1978. See Szasz, 1982:353-355, for a summary of these polls, as well as of other polls that found similar trends.

8. A total of 85 percent approved setting stricter auto emission standards (1978 Cambridge Reports Poll, cited in Szasz, 1982:349), 93 percent favored stricter toxic chemical dumping standards (ABC News/Harris Survey, 1980, cited in ibid.:350); 86 percent favored making the Clean Air Act stricter or keeping it about the same; 93 percent favored making the Clean Water Act stricter or keeping it about the same (1981 Harris Survey, cited in ibid.:350).

9. Within the administration, economists at the OMB, the Council of Economic Advisers, and the Council on Wage and Price Stability repeatedly intervened in the EPA's, OSHA's, and other agencies' rule making in the name of the administration's anti-inflation effort. A Reg-

ulatory Analysis and Review Group was set up to "integrate . . . economic analysis into the early stages of the decision-making process [and] encourage agencies to develop better analysis as an integral part of the regulation writing process" (Bureau of National Affairs, *Occupational Safety and Health Reporter*, April 14, 1977). Regulatory reform bills were introduced in the 95th and 96th Congresses (1977-80). The Supreme Court would rule that regulators had to demonstrate a "reasonable relationship" between the costs and benefits of proposed standards (*Industrial Union Department, AFL-CIO v. American Petroleum Institute*, 1980). The policy side of the deregulatory campaign is described in detail in Szasz, 1982, 1984, 1986c.

10. As we have already seen, above, Stockman honed his antiregulatory rhetoric during congressional consideration of Superfund. Here is one example of Stockman at work: "H.R. 7020, the 'Hazardous Waste Containment Act of 1980,' is a replay of the patented formula for environmental legislation developed over the past decade. A barrage of statutory language employing ill-defined concepts; a vast administrative delegation with no sound analysis of the problem; a lack of explicit policy choices and parameters; and an implicit cost of billions in national economic resources for marginal or non-existent social benefits. Having established this pattern of regulatory overkill in the 1970's, the nation is now paying the price in the form of worsening economic conditions, stagflation, collapsing productivity and international competitiveness, declining living standards, and rising welfare costs" (in Senate, 1983, vol. 2:101).

11. Cited by Bingham, 1981.

12. See, for example, his written statements in Senate, 1983, vol. 2:101-106, and his comments on the floor of the House in ibid.:298-303.

13. For more background on Gorsuch and Lavelle, see Harris et al., 1987:25-34. Typically, the people appointed to be assistant administrators came to EPA having previously worked for corporations or for law firms that had done battle with the federal regulatory agencies. Some of the other figures who were fired in the aftermath of Sewergate, for example, came to the EPA from Johns Manville, the asbestos manufacturer, and from Exxon.

14. See chapter 2, note 54.

15. For detailed documentation of these policies, see National Wildlife Federation, 1982.

16. For a summary of the administration's actions on RCRA, see testimony by the Environmental Defense Fund in House, 1982a:219-232.

17. Fortuna and Lennett, 1987:12-13.

18. Barke, 1988:152; Epstein et al., 1982:250; Fortuna and Lennett, 1987:13-14; Lieber, 1983:64; Senator Hart in Senate, 1982:25-26.

19. See GAO, 1983c, for a summary of RCRA implementation through 1983; some highlights follow. *Groundwater requirements*: In two of the four states visited by GAO, 78 percent of facilities were not in compliance with groundwater monitoring requirements; the other two states had no idea of compliance rate because most facilities had not been inspected. *Closure, postclosure, and financial responsibility*: "None of the four states we reviewed required their inspectors to routinely evaluate in detail the adequacy of a facility's closure and postclosure plans . . . because the states lacked adequate federal guidance on how to perform the evaluations, had limited inspection resources, or had not yet adopted financial responsibility requirements. . . . the extent of nationwide noncompliance is unknown" (ibid.:2). *Inspection and enforcement*: in five states surveyed by GAO, less than half, 45 percent, of facilities had been inspected. "Most violations (75 percent) resulted in the issuance of warning letters or notices of violation. Few compliance orders had been issued, and penalties totaling $142,375 were assessed against nine facilities" (ibid.:3). *Permitting*: "Through July, 1983, only 24 of the estimated 8,000 facilities expected to require permits had received final permits" (ibid.:4).

20. GAO, 1982.

21. In the handful of cases where negotiations with responsible parties were carried to conclusion, EPA tended to reach settlements that were not too tough on the polluting firms.

22. Another GAO report, 1983a, suggested that the little that was being spent was spent badly.

23. GAO, 1985b.

24. By December 31, 1984, well after the post-Sewergate speedup of Superfund activity, the EPA considered cleanup action completed at ten sites on the National Priority List (NPL). The GAO noted dryly, however, that "the 10 sites EPA considered cleaned up generally involved relatively uncomplicated remedies compared to problems EPA currently faces at most NPL sites" (1985d:6).

25. Parts of the discussion in this section are revised from Szasz, 1986a.

26. For complaints about lack of agency cooperation, see Harris et al., 1987:31; Senate, 1982:19. The hostile exchanges between senators and Rita Lavelle in Senate, 1982, show the level to which interaction between congressional liberals and administration appointees had sunk.

27. Harris et al., 1987:28-29.

28. Ibid.:31.

29. See, for example, Grozuczak, 1982; National Wildlife Federation, 1982.

30. These allegations are documented in House, 1985.

31. On this, investigators for the House Judiciary Committee (House, 1985) agreed with Anne (Gorsuch) Burford's congressional testimony in October 1983.

32. Between February 3, when Judge Smith handed down his decision, and February 21, when the administration's attempt to strike a deal with Congress fell through, media coverage expanded dramatically. The *New York Times* had the EPA on the front page on twelve of the nineteen days, and ran thirty-one total stories and four editorials.

33. Between February 23, when the Democrats rejected the compromise offer, and March 9, the day before Anne Gorsuch resigned, the *New York Times* had the EPA on the front page thirteen of fifteen days, and ran a total of forty-six stories, an average of more than three articles a day.

34. See editorials in the *New York Times* February 16 and March 2. On March 6 the *Times* approvingly quoted similar editorials from major dailies in Washington, Philadelphia, Kansas City, Miami, Arizona, Chicago, and Detroit, and the *Wall Street Journal*.

35. Washington Post/ABC Poll, March 5, 1983.

36. One can argue the case for administration miscalculation. Thus, scandal occurred because of blunders that resurrected the loaded symbolism of Watergate—paper shredders, executive privilege, coverup. Undoubtedly, this played a part in shaping the development of the scandal. However, it still does not explain what emboldened Congress in the first place, what motivated it to begin pushing and keep pushing until the administration blundered. That original thing was hazardous waste.

37. Szasz, 1982, ch. 6.

38. It would be wrong to argue that the dismissal of James Watt, secretary of interior, was the first big victory for antideregulatory forces. Certainly, he was a major target, both for his specific policies and, more generally, as the symbol of the administration's position on development versus environment. It should be remembered, however, that Watt was ultimately brought down by blunders having little to do with his environmental positions, especially by controversy following his wisecrack about a commission to which he had appointed "a Black, . . . a woman, two Jews and a cripple."

39. Alm, 1984:1; Shabecoff, 1983.

40. Ruckelshaus, 1984a; 1984b:287.

41. Ruckelshaus, 1984a, 1984b; Shabecoff, 1983. The GAO (1985b:6, 7) confirmed that the pace of cleanup activity did speed up after Sewergate.

42. GAO, 1985d:3.

43. Bowman, 1988:137.

44. EPA and GAO used somewhat different criteria for declaring a site completed.

45. GAO, 1988c.

46. RCRA's original regulatory scheme had been based on the assumption that, in theory, competent regulation could make land disposal of hazardous wastes an environmentally sound practice; several new studies showed conclusively that landfills always eventually fail and inevitably pollute groundwater, threatening public health and the environment. See *RACHEL*, 71, 109, 116, 117, 119, 125. It was now also generally understood that some of the loopholes in the original act—exemptions for firms generating less than 1,000 kilograms (2,200 pounds) of hazardous waste a month, exemptions for wastes blended with oil and burned in boilers, and so on—left significant waste streams unregulated. Everyone was abundantly aware of the agency's chronic inability and/or unwillingness to implement even this seriously flawed design.

47. Fortuna and Lennett, 1987:14-15.

48. Ibid.; Harris et al., 1987:31; House, 1982a; Senate, 1982. On the other side, only the Reagan administration and chemical manufacturers argued that things were working just fine. Rita Lavelle, then assistant administrator at the EPA, told one committee, "We believe that most wastes can be satisfactorily managed in the land and that it can be done with a reasonable margin of safety more cheaply in this manner. . . . it may be that recycling or destruction is preferable from a strictly health and environmental protection standpoint, but for many wastes, the reduction in risk achieved is probably marginal and may not be worth the cost" (quoted in Harris et al., 1987:xvii). The Chemical Manufacturers Association testified that Congress should "continue to allow appropriate land disposal of hazardous wastes," that "further amendment of the statute is not now appropriate" (House, 1982a:314, 316).

49. Congress's failure to act on RCRA was entirely typical of its behavior during this period. Other environmental laws were also due for reauthorization—the Clean Air and Clean Water Acts, for example—but "the political consensus necessary to actually pass a major new piece of environmental legislation was apparently missing; . . . [Other than the post-Sewergate reauthorization of RCRA, no] major piece of environmental legislation emerge[d] from the four years of the 97th and 98th Congresses. Since the enactment of the National Environmental Policy Act in 1969, . . . no comparable hiatus of congressional action in the environmental arena has occurred" (Mugdan and Adler, 1985:215).

50. As J. William Futrell, president of the Environmental Law Institute, commented, "The headline-grabbing controversies that characterized the Environmental Protection Agency in late 1982 and early 1983 provided that tablet on which the RCRA Amendments were to be written" (Harris et al., 1987:viii). Congressman Eckart noted, "This 'reign of error' at EPA provided considerable momentum toward enactment of a stringent RCRA reauthorization bill" (ibid.:xviii).

In the literature, one finds two other, competing, explanations for why Congress found its courage and moved to strengthen RCRA: (1) "Policy learning," that is, new information about the problem, led members of Congress to see that their first legislative effort was insufficient and would have to be revised; and (2) the pluralist argument, that Congress acted because of the wide coalition of interests—scientific bodies, state government officials, environmental organizations, the hazardous waste industry—that called for a fundamental overhaul of RCRA. Undoubtedly, both were contributing factors. However, neither would have been sufficient without the presence of the icon-induced, icon-confirming scandal. Policy learning, by itself, is not a plausible explanation. Congress is not a pure, rational problem solver. It does not legislate solely on the basis of better information. Knowledge about some policy's inadequacies becomes salient only in the context of successful attempts to draw attention to those inadequacies. The problems of RCRA were known *before* 1983 and did not, in the absence of scandal, prove to be motivation enough to make reauthorization a success. Similarly, the breadth of the interest groups calling for reform of RCRA proved to be important *after* Sewergate; the same forces were there earlier, criticizing the Reagan administration in congressional oversight hearings, asking for changes. Policy learning and the presence of a broad coalition of interest groups both contributed to reauthorization, but only after the scandal had fundamentally altered the political configuration in Washington and confirmed the hazardous waste issue's cachet.

51. "Almost every section of the RCRA Amendments might be read as expressing a sense of frustration over the pace and scope of EPA action" (Mugdan and Adler, 1985:217).

52. See House, 1983, vol. 1:116-119.

53. Readers interested can consult issues of the *Congressional Quarterly Weekly Report*, September 11, 1982; January 29, March 19, April 30, May 14, June 18, July 30, August 20, and November 5, 1983; July 28 and September 29, 1984.

54. As Representative Florio put it, "What we're doing now is mandating things [limiting] the discretion that's available to the agency, because things that we assumed in a good faith way would be done have not been done" (*Congressional Quarterly Weekly Report*, March 17, 1984:620).

55. *Congressional Quarterly Weekly Report*, March 2, 1985:410.

56. The two sides manuevered, using every tactical device at their disposal. The bills had to work their way through a maze of committees that claimed to have jurisdiction over some part of the legislation. Some things — notably the most far-reaching victim rights provisions — were dropped again, mostly without much of a fight. The liberal position on cleanup standards and timetables prevailed. The funding mechanism — that is, deciding *who would pay* — proved by far the most controversial issue. Various sectors of American industry struggled mightily to have someone else pick up the tab. Oil and chemical companies wanted out of the feedstock tax scheme and advocated, instead, that the money should come from a waste-end tax levied on all generators or from general revenues. Other business associations fought equally hard against business taxes that might fall on them. See especially *Congressional Quarterly Weekly Report*, October 5 and 19, 1985. Readers interested in more detail than provided by the *Congressional Quarterly Weekly Report* can consult the legislative history in Senate, 1990.

57. Provisions of the Superfund Amendments and Reauthorization Act are summarized in Hedeman et al., 1987. Tarlock and Glicksman, 1987, discuss SARA's citizen rights, participation, and information provisions.

58. *Congressional Quarterly Weekly Report*, September 21, 1985:1894.

59. Senate, 1990, vol. 1:171.

7. Fifteen Years of Hazardous Waste Legislation: Summing Up the Policy Impacts

1. Acton, 1989; Greer et al., 1988; OTA, 1988, 1989.

2. This information is available directly from the government, from the EPA, or on Tox-Net, the National Library of Medicine's on-line information service. The activist community analyzes government-provided information for its members in newsletters such as *Ecological Illness, Toxics Watchdog, Waste Not*, and, of course, in RACHEL, the on-line information service. The information is collected, analyzed, reorganized, and sold to lawyers and to industry. See, for example, Environmental Data Resources, 1991.

3. Krag, 1985.

4. Goldman, Hulme, and Johnson, 1986; GAO, 1987b, 1988a, 1988b, 1988c. Movement organizations have made a cottage industry of whistle-blowing on the biggest of the big, Waste Management, Inc. (WMI), and Browning-Ferris Industries (BFI): Greenpeace has a very long file of WMI's violations on its electronic bulletin board, Econet. For stories on both BFI and WMI, see *RACHEL*, 27, 28, 34, 47, 51, 66, 91, 93.

5. See CCHW, 1987b; Freeman, 1987; various issues of *Hazardous Waste and Hazardous Materials*; report on EPA's Superfund Innovative Technology Evaluation (SITE) program in *RACHEL*, 150.

6. Committee on Institutional Considerations in Reducing the Generation of Hazardous Industrial Wastes, 1985.

7. EPA, 1986; OTA, 1986, 1987.

8. Harris et al., 1987:163-165.

9. GAO, 1988c:54.

10. Environmental Defense Fund, 1986; Lennett and Greer, 1985.

11. See *RACHEL*, 71, 109, 116, 117, 119, 125.

12. Congress put generators on the honor system: "The determination of 'economically practicable' will be made by the generator and is not subject to subsequent re-evaluation" (Senate report cited in Harris et al., 1987:163). The waste reduction bills Congress considered subsequently limited federal intervention to setting up a new office at the EPA, providing grants, and creating an information clearinghouse. Generators would be required only to file reports on their reduction efforts. See *Environment Reporter*, 1988, 19:175.

13. These included the NRC conference in 1985 and Government Institute's series of seminars on waste minimization in 1986 (Editor, *Hazardous Waste and Hazardous Materials*, 1985c, 1986). Waste minimization was described as "imperative" at the 1987 meetings of the Air Pollution Control and Hazardous Waste Management Association (*Environment Reporter*, 1987, 18:742), and in 1988, a waste minimization conference was sponsored by the Environment and Energy Study Institute (*Environment Reporter*, 1988, 19:38).

14. *Environment Reporter*, 18:742, 1987; Editor, *Hazardous Waste and Hazardous Materials*, 1985c, 1986; Huggins and Towery, 1987; Johnnie, 1987; Petros, 1987; Purcell, 1986.

15. "By claiming that waste reduction efforts are at this point good business, the industry officials were saying no regulatory stick is necessary" (report on the 1987 Air Pollution Control and Hazardous Waste Management Association meetings, *Environment Reporter*, 1987, 18:742).

16. Committee on Institutional Considerations in Reducing the Generation of Hazardous Industrial Wastes, 1985; Oldenburg and Hirschhorn, 1987; EPA, 1986; Editor, *Hazardous Waste and Hazardous Materials*, 1986; EPA study by ICF, Inc., of Fairfax, Virginia (see *RACHEL*, 81); McCoy and Associates, 1986:4-4.

17. Clymer, 1989.

18. See House, 1982a, 1982b; Senate, 1983.

19. Environmental law is a growth industry in corporate law. See Macbeth, n.d.; Rutter, 1991. Superfund litigation now constitutes a major share of the environmental practice at big law firms. A comprehensive legal/information infrastructure is in place, including newsletters (e.g., *Hazardous Waste Litigation Reporter*), computerized searches of regulatory data on industrial sites (Environmental Data Resources' HAZ-SITE Reports), seminars and conferences (CLE International, "Hazardous Waste in Real Estate Transactions," 1990; Professional Education Systems, "Environmental Liability in Real Estate Transactions in California," 1990). When industrial property changes hands, the new owner is strongly motivated (1) to establish what wastes have been disposed at the site, and (2) to negotiate responsibility for cleanup with the previous owner as a condition of the sale. In this way, regulation is, in part, implemented by private parties instead of the regulators, thereby improving our overall evaluation of the quality of implementation.

20. *RACHEL*, 5. See also Ottinger, 1985:20-21.

21. My argument is that this is an *objective* result that is not necessarily subjectively intended by participants in the movement. One can certainly find the occasional movement leader or theorist who *consciously* intended to block siting in order to force the system toward waste reduction. In an interview conducted for this volume, John Thompson of CSRC said that "in 1982, when we started, our model was sort of to constipate the system. We felt there would be no demand for recycling as long as there was a cheap source of land disposal. By knocking out the sites . . . we felt we would drive up the costs enough that other options such as recycling would more likely be done." However, my argument does not depend on member

consciousness or intentionality. As I say, the scissor effect is an objective outcome, the effect of the sheer aggregate impact of local refusals, regardless of the subjective understanding or conscious intention of the participants.

I must also note that waste reduction is only one possible response to the kinds of legal and monetary pressures described here. There have been other, less socially desirable, reactions as well. Some generators have reacted by exporting wastes to developing nations. See, for example, Brooke, 1988. Avoiding regulation by moving one's operations or one's banned products to other nations is common. It should alert us to the need, not sufficiently emphasized so far in this book, that environmental problems and solutions must of necessity now be couched in global, not national, terms.

22. Reflection on the connection between grass-roots action and industry's "voluntary" movement toward waste reduction can be pursued in two somewhat different directions. One can emphasize its historical specificity, its role in forcing society to take up the shift toward source reduction, as I have done in the main text. Or one can attempt, instead, to draw generalizable tactical lessons from the experience.

In chapter 2, I discussed the historical and structural fact, the state-economy relationship that severely limits the political system's capacity or willingness to direct economic activity for the sake of collective, societal goals. As a result, even when it is abundantly clear that disposal regulation is doomed to chronic failure and everyone agrees that waste reduction is the only real solution, policymakers twist this way and that, willing to educate, facilitate, encourage, and cajole, but never, never to dictate waste reduction.

The alternatives do not seem particularly promising: on the one hand, effective policy that cannot be adopted because it directly transgresses one of the core defining exclusion zones of the political in capitalist society; on the other hand, policy that never accomplishes more than a fraction of its goal. Is there another option, other than either waiting for ecosocialism or some such utopia or living with the currently available extremes of "successful, but impossible" and "possible, but futile"?

My analysis of the "scissor" effect created by the hazardous waste movement suggests that there *is* another option. Even if the state refuses to intervene directly in production to force pollution reduction, similar results can be achieved indirectly and incrementally when social movements use the democratic mechanisms of the political system, at all levels simultaneously, to drive up economic costs and force imposition of stronger legal liabilities, thereby creating incentives that generators understand.

The anti-nuclear power movement appears to have achieved its goals with a similar pattern of action. Opponents of nuclear power fought for more stringent regulatory requirements as well as for procedural safeguards such as greater public participation in the licensing process. Stronger regulations drove up the cost of building and running nuclear power plants. Opponents used the public intervenor provisions to cause delays, to insist on even more safeguards, finally to drive up costs to the point that utilities found nuclear energy uneconomical.

Generalizing, this appears to be one promising way forward at this moment in history. The way to transgress the historical/structural limits that have given us the failed model of regulation is for social movements to *use and abuse simultaneously* the democratic aspects of the state, to fight, on the one hand, to strengthen regulations (including participation provisions) and, on the other, to mobilize grass-roots movements and environmental activist organizations to use those regulatory provisions in a way that disables implementation, drives up costs, and generally interferes with the commodification of nature at every turn. Elsewhere, I have dubbed this kind of social action "policy Luddism" (Szasz, 1992).

23. Starr, 1969.

24. For various positions on the hazards of hazardous waste facilities, ranging from assertions of definite risk, through carefully articulated positions of uncertainty, to flat out denials that there is much of anything to fear, see Goldman et al., 1985; Grisham, 1986; Harris,

Highland, Rodericks, and Papadopulos, 1984; Kolata, 1980; Lowrance, 1981; Majumdar and Miller, 1984; Maugh, 1982; Paigen et al., 1985; GAO, 1978b.

25. Goldman et al., 1986.

26. To cite just a few: Ohio EPA "ordered CECOS OH to halt construction of hazardous waste cell #8, because ground water pressure collapsed a portion of the cell wall. . . . U.S. EPA suspended CECOS OH's TSCA permit for PCB disposal from February 22 to April 13 [1983] as a result of the slope failure. . . . On November 9 OEPA shut down the facility after receiving reports that its operators pumped phenol-contaminated water from a landfill cell into a tributary that joins the East Fork of the Little Miami River upstream of public drinking water intakes. . . . indications of ground water contamination were detected [in 1985] . . . the newly constructed cell #9 did not meet minimum technology standards. OEPA closed the facility again pending development of an adequate ground water monitoring program. In June a county grand jury indicted BFI, CECOS, and two former site employees on 96 counts. The criminal indictment alleges that site employees deliberately pumped contaminated rainwater into a nearby creek" (ibid.:239).

27. See note 11.

28. An incinerator owned by Rollins Environmental Services was shut down by the Louisiana Department of Environmental Quality for violating air quality regulations. The inspector found "black smoke at 100 percent opacity being emitted from the incinerator's stack. . . . A carbon monoxide monitor was observed to be malfunctioning, and the control room was being operated in a confused manner. . . . 'the control room operator seemed ready to faint' " (*Environment Reporter*, September 6, 1985:816). This facility, which had a deep injection well, landfill, and chemical and biological treatment impoundments in addition to the incinerator, had been cited for air and water pollution violations eight times in five years, and the state regulator declared her intent to shut down the whole facility permanently (ibid.).

Rollins operated a third of the major commercial hazardous waste incinerators in the United States in 1985. To put things in proper perspective, it should be noted that Rollins was one of the big eight firms in the CEP study. CEP rated Rollins *better*, overall, than either CECOS (a Browning-Ferris Industries subsidiary) or Chemical Waste Management (the Waste Management, Inc., subsidiary), the two largest firms in hazardous waste management (Goldman et al., 1986).

When the House of Representatives held hearings on hazardous waste disposal in 1981, the landfill at Wilsonville, Illinois, was described, by Dr. Raymond D. Harbison, toxicologist, EPA consultant, and professor of pharmacology at Vanderbilt University, as "the most scientific landfill in this country" (House, 1981a:267). Geological and soil permeability feasibility tests were conducted to verify the suitability of the site before construction was begun. Yet subsequent studies showed that the soil was more porous than originally thought and water was found seeping in at rates greater than predicted. Furthermore, the landfill was built over an abandoned coal mine, and feasibility tests underestimated the likelihood of subsidence.

George J. Tyler, assistant commissioner of the New Jersey Department of Environmental Protection, said about the Lone Pine landfill in Freehold, New Jersey, "The landfill is leaking into the water, but so does every landfill in this country" (House, 1981b:188).

8. Broader Political Implications? Environmental Populism and the Reconstitution of Progressive Politics

1. CCHW's songbook (1989) encourages members to sing traditional union songs, such as "Roll the Union On." See also Lois Gibbs's praise of the labor movement in her essay in CCHW, 1989.

2. CCHW also sings civil rights songs. Gibbs's essay in CCHW, 1989, pays homage to Frederick Douglass, Sojourner Truth, Fanny Lou Hamer, and Ella Baker.

3. CCHW, 1986b.

4. The women leaders of grass-roots groups are praised as the contemporary sisters of the "long chain of great women who have been leaders in the struggle for social justice." Sojourner Truth, Elizabeth Cady Stanton, Mother Jones, Ella Baker, Fanny Lou Hamer, Jessie Lopez de la Cruz, and Delores Huerta, among others, are mentioned. "These women, and thousands and thousands of others whose names have been lost, were American leaders for social change. . . . They are with us in spirit. Our story is their story" (Zeff et al., 1989:1).

5. Lois Gibbs: "When an incinerator was being sited, or whatever it might have been, one of the questions that we often ask . . . is: 'Why is it being sited here? What's unique about this area?' And they [the local citizens] would realize that 'It's lower income,' or 'We have higher unemployment,' and go through a list of things that make it attractive to industry because it makes it, at least theoretically, a politically weak city. . . . [When, then, people asked,] 'If the root of the problem is that we're economically weak, . . . how can we change that?' . . . One of the things that comes up is economic development, sound economic development" (interview).

6. Kaye Kiker: "ACE has about 350 members and we are biracial. That was one mountain we had to climb. A lot of people left ACE because we decided to be a biracial group, but we've gotten beyond that" (quoted in CCHW, *Everyone's Backyard*, 1989, 7[4]:4). Sue Greer of PAHLS understands that organizing around hazardous waste means organizing every kind of powerless and disadvantaged American: "I think that it is unfair that the rich dump on the poor. And that's what this issue is all about. . . . Hey, if you're poor and you're black and you're Hispanic or you're a farmer, like me, look out 'cause here comes an incinerator or landfill or some God-awful piece of pollution." But when her organization went to Gary, Indiana, to a "very low income, minority area . . . very depressed . . . lots of laid-off steel workers . . . predominantly black," to organize against a proposed toxic waste incinerator, "the black mayor of Gary and the City Council told us to go home, we didn't have any business over there, although it's like fifteen miles from where I live. It's the same air. Air isn't black or white, we all breathe the same air. They tried to make a racial issue out of it, I think. . . . I found that there's two classes of blacks in the Gary area. There are some very racist people over there. It's very hard to work with them. But there are some very beautiful people over there, too, and those are the people that we are bringing in now" (interview).

"If you want to win your local environmental fight in a multi-racial community, it is essential to actively recruit and welcome people of color into your group. . . . Industry is often accused of trying to divide local environmental groups in order to make them less effective . . . but local groups may more effectively divide and conquer themselves if they remain predominantly white. . . . In the long run, what is the future or the Grassroots Movement for Environmental Justice if it isn't a broad-based, multi-racial movement? . . . *we're* going to have to show some foresight, by actively working to overcome racial, cultural and economic obstacles" (Carter, 1990; see also Ruffins, 1990).

7. For critiques of traditional environmentalism's relationship to issues of race and inequality, see Anthony, 1990; Bullard and Wright, 1990; Lewis, 1990.

8. As discussed in chapter 4, threats to family health are the biggest factor in citizen activism. "Everything is a women's issue because every child that's born, some woman had it" (Cora Tucker, quoted in Zeff et al., 1989:5; see also Hamilton, 1990). As a result, "the grassroots movement against toxics is a movement dominated and led by women. 70%-80% of the local leaders . . . are women" (Zeff et al., 1989:25).

9. Zeff et al., 1989:27.

10. This last paragraph is from a Greer article in *PAHLS*, 1988, 6(3). The rest comes from Hal Aronson's interview with Greer for this volume.

11. Marty Chestnutt: "I think they get a lot from the newsletters. Also, once you are on the newsletter mailing list, you get onto other mailing lists from other progressive organizations and pretty soon you are getting information from all over the U.S. My mailman flips every time

I get the socialist newspaper, which is something I didn't subscribe to. . . . I read it and I get a lot of good ideas out of it" (interview).

12. Hal Aronson asked Lois Gibbs, "Do you think this change in ideology is happening to a lot of the people who are locally involved, or do you think it is happening to a few of them?" Her reply, "It's happening to a lot of them at very different levels and very different scales, but most of them have moved beyond wherever they were within a year after they begin their fight."

References

Acton, Jan Paul. 1989. *Understanding Superfund: A Progress Report*. Santa Monica, CA: RAND Corporation.

Albany State College Toxics Communications and Assistance Project. 1989. *Southern Action*. Albany, GA: Albany State College.

Alm, Alvin. 1984. "Opening Address," pp. 1-4 in ABA Standing Committee on Environmental Law, *Siting of Hazardous Waste Facilities and Transport of Hazardous Substances*. Washington, DC: Public Services Division, American Bar Association.

Anderson, Richard F., and Michael R. Greenberg. 1984. "Siting Hazardous Waste Management Facility: Theory versus Reality," pp. 170-186 in Majumdar and Miller, 1984.

Andrews, Richard N.L. 1988. "Hazardous Waste Facility Siting: State Approaches," pp. 117-128 in Davis and Lester, 1988.

Anonymous, *Glamour*. 1980a. "Tell Us What You Think About . . . Love Canal." August, p. 42.

_____. 1980b. "What You Thought About . . . Love Canal." November, p. 31.

Anonymous, *Time*. 1978. "A Nightmare in Niagara." August 14, p. 46.

_____. 1980a. "Explosion of a Toxic Time Bomb: New Jersey Blaze Spotlights Growing Problem of Chemical Waste." May 5, p. 67.

_____. 1980b. "The Neighborhood of Fear: Carter Orders a Further Evacuation from Polluted Love Canal." June 2, pp. 61-62.

_____. 1980c. "Skeptical View: Another Look at Love Canal." November 3, p. 99.

Anthony, Carl. 1990. "Why African Americans Should Be Environmentalists." *Race, Poverty and the Environment* 1(1):5-6.

Ashford, Nicolas A. 1976. *Crisis in the Workplace*. Cambridge: MIT Press.

Associated Press. 1989. "Alabama Bans Many Toxic Waste Shipments." *New York Times*, August 31.

Bacow, Lawrence S., and James R. Milkey. 1987. "Overcoming Local Opposition to Hazardous Waste Facilities: The Massachusetts Approach," pp. 159-205 in Lake, 1987.

Baran, Paul A. 1957. *The Political Economy of Growth*. New York: Monthly Review Press.

Baran, Paul A., and Paul M. Sweezy. 1966. *Monopoly Capital: An Essay on the American Economic and Social Order*. New York: Monthly Review Press.

Barke, Richard. 1988. "Hazardous Wastes and the Politics of Policy Change," pp. 147-162 in Davis and Lester, 1988.

Baudrillard, Jean. 1983. "In the Shadow of the Silent Majorities," pp. 1-61 in *In the Shadow of the Silent Majorities, Or the End of the Social, and Other Essays*. New York: Semiotext(e).

Beck, Melinda, and Mary Lord. 1979. "A Caustic Report on Chemical Dumps." *Newsweek*, October 22, p. 51.

Beck, Melinda, Mary Lord, and Jerry Buckley. 1979. "More Love Canals?" *Newsweek*, May 14, p. 41.

Begley, Sharon. 1980. "Toxic Waste Still Pollutes Roadways." *Newsweek*, October 27, pp. 25-25A.

Bennett, W. Lance. 1988. *News: The Politics of Illusion* (2nd ed). New York: Longman.

Berman, Daniel M. 1978. *Death on the Job*. New York: Monthly Review Press.

Bernstein, Marver H. 1955. *Regulating Business by Independent Commission*. Princeton, NJ: Princeton University Press.

Bhatt, Harasiddhiprasad, Robert M. Sykes, and Thomas L. Sweeney, eds. 1985. *Management of Toxic and Hazardous Wastes*. Chelsea, MI: Lewis.

Bingham, Eula. 1981. [Letter to the editor, business section]. *New York Times*, February 1.

Bingham, Gail. 1984. "Prospects for Negotiation of Hazardous Waste Siting Disputes," pp. 17-19 in ABA Standing Committee on Environmental Law, *Siting of Hazardous Waste Facilities and Transport of Hazardous Substances*. Washington, DC: Public Services Division, American Bar Association.

Block, Fred. 1984. "The Ruling Class Does Not Rule: Notes on the Marxist Theory of the State," pp. 32-46 in Thomas Ferguson and Joel Rogers, eds., *The Political Economy: Readings in the Politics and Economics of American Public Policy*. Armonk, NY: M. E. Sharpe.

Blumenthal, Ralph. 1983a. "Toxic Dumping Unit in Jersey Is Rebuked." *New York Times*, January 16.

———. 1983b. "Corruption Cited in Waste Business." *New York Times*, May 1.

Bowman, Ann O. 1988. "Superfund Implementation: Five Years and How Many Cleanups?" pp. 129-144 in Davis and Lester, 1988.

Bowman, Ann O., and James P. Lester. 1985. "Hazardous Waste Management: State Government Activity or Passivity?" *State and Local Government Review* 17(1):155-161.

Brodeur, Paul. 1973. *Expendable Americans*. New York: Viking.

Brooke, James. 1988. "Waste Dumpers Turning to West Africa." *New York Times*, July 17.

Brown, Michael. 1979. *Laying Waste: The Poisoning of America by Toxic Chemicals*. New York: Pantheon.

Brown, Phil, and Edwin J. Mikkelsen. 1990. *No Safe Place: Toxic Waste, Leukemia, and Community Action*. Berkeley: University of California Press.

Bryant, Bunyan, and Paul Mohai, eds. 1992. *Race and the Incidence of Environmental Hazards: A Time for Discourse*. Boulder, CO: Westview.

Buckingham, Phillip L., Ronald J. Buchanan, and Michael D. LaGrega. 1986. "The Strategies and Choices in Pennsylvania's Hazardous Waste Facilities Plan," pp. 489-501 in Gregory D. Boardman, ed., *Toxic and Hazardous Wastes: Proceedings of the Eighteenth Mid-Atlantic Industrial Waste Conference*. Lancaster, PA: Technomic.

Bullard, Robert D. 1990. *Dumping on Dixie*. Boulder, CO: Westview.

Bullard, Robert D., and Beverly H. Wright. 1990. "The Quest for Environmental Equity: Mobilizing the Black Community for Social Change." *Race, Poverty and the Environment* 1(2):3, 14-17.

Bureau of National Affairs, Inc. *Environment Reporter*, various issues.

———. *Occupational Safety and Health Reporter*, various issues.

Bush, Paul D. 1981. "Citizen Response to Industrial Waste," pp. 533-536 in Huang, 1981.

Callinicos, Alex. 1989. *Against Postmodernism: A Marxist Critique*. New York: St. Martin's.

Canter, Bram D. E. 1981. "Safe Hazardous Waste Disposal: Sure, but Where?" *Florida Bar Journal* 55:813-817.

———. 1982. "Hazardous Waste Disposal and the New State Siting Programs." *Natural Resources Lawyer* 14(3):421-456.

Carnes, Sam A. 1982. "Confronting Complexity and Uncertainty: Implementation of Hazardous Waste Management Policy," pp. 35-50 in Dean E. Mann, ed., *Environmental Policy Implementation*. Lexington, MA: Lexington.

Carson, Rachel. 1962. *Silent Spring*. Boston: Houghton Mifflin.

Carter, Clay. 1990. "Race and Environmental Organizing: Pulling Down the Barriers." *Everyone's Backyard* 8(2):6-8.

Cerrell Associates and J. Stephen Powell. 1984. *Political Difficulties Facing Waste-to-Energy Conversion Plant Siting*. Sacramento: California Waste Management Board.

Chemical Manufacturers Association. n.d. *A Statute for the Siting, Construction and Financing of Hazardous Waste Treatment Disposal and Storage Facilities*. Washington, DC: Author.

Chesler, Robert D., Michael L. Rodburg, and Cornelius C. Smith, Jr. 1986. "Patterns of Judicial Interpretation of Insurance Coverage for Hazardous Waste Site Liability." *Rutgers Law Journal* 18(1):9-72.

Citizen's Clearinghouse for Hazardous Wastes, Inc. 1984. *Annual Report, 1984*. Arlington, VA: Author.

_____. 1985. *Annual Report, 1985*. Arlington, VA: Author.

_____. 1986a. *Annual Report, 1986*. Arlington, VA: Author.

_____. 1986b. *Five Years of Progress, 1981-6*. Arlington, VA: Author.

_____. 1987a. *Annual Report, 1987*. Arlington, VA: Author.

_____. 1987b. *Advanced Treatment Technologies for Disposal of Hazardous Wastes*. Arlington, VA: Author.

_____. 1988. *Annual Report, 1988*. Arlington, VA: Author.

_____. 1989. *Grassroots Convention '89 Songbook*. Arlington, VA: Author.

_____. *Everyone's Backyard*, various issues.

_____. *Action Bulletin*, various issues.

Clapham, W. B., Jr. 1985. "Conflicts and Hazardous Waste Management: The Environmental Viewpoint," pp. 9-18 in Bhatt, Sykes, and Sweeney, 1985.

Clark, Matt, Mary Hager, Dan Shapiro, and William Marbach. 1980. "Fleeing the Love Canal." *Newsweek*, June 2, pp. 56-57.

Claybrook, Joan, and the Staff of Public Citizen. 1984. *Retreat from Safety: Reagan's Attack on America's Health*. New York: Pantheon.

Clines, Francis X. 1983. "White House Seeks to Peer Beyond E.P.A. Smoke." *New York Times*, March 11.

Clymer, Adam. 1989. "Polls Contrast U.S.'s and Public's Views." *New York Times*, May 22.

Cobb, Roger W., and Charles D. Elder. 1983. *Participation in American Politics: The Dynamics of Agenda-Building* (2nd ed.). Baltimore: Johns Hopkins University Press.

Cohen, Steven, and Marc Tipermas. 1983. "Superfund: Preimplementation Planning and Bureaucratic Politics," pp. 43-59 in Lester and Bowman, 1983.

Commission for Racial Justice, United Church of Christ. 1987. *Toxic Wastes and Race in the U.S.: A National Report on the Racial and Socio-Economic Characteristics of Communities with Hazardous Waste Sites*. Cleveland, OH: Author.

Committee on Institutional Considerations in Reducing the Generation of Hazardous Industrial Wastes, Environmental Studies Board, Commission on Physical Sciences, Mathematics, and Resources, National Research Council. 1985. *Reducing Hazardous Waste Generation: An Evaluation and a Call for Action*. Washington, DC: National Academy Press.

Commoner, Barry. 1989. "Why We Have Failed." *Greenpeace* 14 (September/October):12-13.

Conference Board. 1980. *Regulatory Problems and Regulatory Reform: The Perceptions of Business* (Report 769). New York: Author.

Converse, Philip E., Jean D. Dotson, Wendy J. Hoag, and William H. McGee III. 1980. *American Social Attitude Data Sourcebook, 1947-1978*. Cambridge, MA: Harvard University Press.

Costle, Douglas M., and Eckardt C. Beck. 1980. "Attack on Hazardous Waste: Turning Back the Toxic Tide." *Capital University Law Review* 9(3):425-433.

Craig, Robert W., and Terry R. Lash. 1984. "Siting Nonradioactive Hazardous Waste Facilities," pp. 99-110 in Harthill, 1984.

Darnovsky, Marcy. 1990. "Social Movements in the Media Environment." Unpublished qualifying essay, History of Consciousness, University of California, Santa Cruz.

Davis, Charles E. and James P. Lester, eds. 1988. *Dimensions of Hazardous Waste Politics and Policy*. New York: Greenwood.

Debord, Guy. 1977. *Society of the Spectacle*. Detroit: Black & Red.

Dobb, Maurice. 1963. *Studies in the Development of Capitalism*. New York: International.

Dodd, Frank J. 1986. "Siting Hazardous Waste Facilities in New Jersey: Keeping the Debate Open." *Seton Hall Legislative Journal* 9(2):423-436.

Downs, Anthony. 1967. *Inside Bureaucracy*. Boston: Little, Brown.

_____. 1972. "Up and Down with Ecology: The 'Issue-Attention Cycle.' " *The Public Interest* 28(Summer):38-50.

Duberg, J. A., M. L. Frankel, and C. M. Niemczewski. 1983-84. "Siting of Hazardous Waste Management Facilities and Public Opposition." *Environmental Impact Assessment Review* 1(1):84-88.

Dunlap, Riley E. 1986. "Two Decades of Public Concern for Environmental Quality: Up, Down and Up Again." Paper presented at the annual meeting of the American Sociological Association, New York.

_____. 1987. "Polls, Pollution, and Politics Revisited: Public Opinion on the Environment in the Reagan Era." *Environment* 29 (July/August):6-11, 32-37.

Earth Island Institute. *Race, Poverty and the Environment*, various issues.

Eckman, Fern Marja. 1980. "Our Fear Never Ends." *McCall's*, June, pp. 94-95, 134ff.

Edelman, Murray. 1964. *The Symbolic Uses of Politics*. Urbana: University of Illinois Press.

_____. 1971. *Politics as Symbolic Action: Mass Arousal and Quiescence*. New York: Academic Press.

_____. 1977. *Political Language: Words That Succeed and Policies That Fail*. New York: Academic Press.

_____. 1988. *Constructing the Political Spectacle*. Chicago: University of Chicago Press.

Edelstein, Michael R. 1988. *Contaminated Communities*. Boulder, CO: Westview.

Editor, *Hazardous Waste and Hazardous Materials*. 1985a. "Editorial," *Hazardous Waste and Hazardous Materials* 2(1):n.p.

_____. 1985b. "Editorial," *Hazardous Waste and Hazardous Materials* 2(2):n.p.

_____. 1985c. "News and Comment," *Hazardous Waste and Hazardous Materials* 2(4):viii.

_____. 1986. "News and Comment," *Hazardous Waste and Hazardous Materials* 3(3):n.p.

_____. 1987. "Editorial," *Hazardous Waste and Hazardous Materials* 4(1):vii.

_____. 1988. "Editorial," *Hazardous Waste and Hazardous Materials* 5(1):ix.

Editors, *Business Today*. 1977. "BT Interview: William Sneath, Chairman, Union Carbide." *Business Today* 14(1):9-13.

_____. 1978. "BT Interview: Irving S. Shapiro, DuPont." *Business Today* 15(1):45-48.

Editors, *Southern Exposure*. 1981a. " 'Get in There and Fight': Taking on Toxic Dumping." *Southern Exposure* 9(3):26-27.

_____. 1981b. " 'Together We Can Do It': Fighting Toxic Hazards in Tennessee; Interview with Nell Grantham." *Southern Exposure* 9(3):42-47.

Elliott, Michael L. Poirier. 1984. "Improving Community Acceptance of Hazardous Waste Facilities Through Alternative Systems for Mitigating and Managing Risk." *Hazardous Waste* 1(3):397-410.

Engels, Friedrich. 1845. *The Condition of the Working-Class in England*. Moscow: Progress.

Environmental Data Resources. 1991. *DATAlink* 1(1).

Environmental Defense Fund. n.d. *EDF Fact Sheet: Closure Plan Guide*. Washington, DC: EDF/Environmental Information Exchange.

_____. 1986. *Approaches to Source Reduction of Hazardous Waste: Practical Guidance from Existing Policies and Programs*. Claremont: California Institute of Public Affairs.

Environmental Research Foundation. *RACHEL's Hazardous Waste News*, various issues.

Epstein, Samuel S., Lester O. Brown, and Carl Pope. 1982. *Hazardous Waste in America*. San Francisco: Sierra Club Books.

Erikson, Robert S., Norman R. Luttbeg, and Kent L. Tedin. 1980. *American Public Opinion* (2nd ed.). New York: Wiley.

Ethridge, Marcus E. 1987. "Procedures for Citizen Involvement in Environmental Policy: An Assessment of Policy Effects," pp. 115-132 in Jack DeSario and Stuart Langton, eds., *Citizen Participation in Public Decision Making*. New York: Greenwood.

Ewen, Stuart. 1976. *Captains of Consciousness: Advertising and the Social Roots of the Consumer Culture*. New York: McGraw-Hill.

Farkas, Alan L. 1980. "Overcoming Public Opposition to the Establishment of New Hazardous Waste Disposal Sites." *Capital University Law Review* 9:451-465.

Forcade, Bill S. 1984. "Public Participation in Siting," pp. 111-122 in Harthill, 1984.

Fortuna, Richard C., and David J. Lennett. 1987. *Hazardous Waste Regulation, the New Era: An Analysis and Guide to RCRA and the 1984 Amendments*. New York: McGraw-Hill.

Fowlkes, Martha R., and Patricia Y. Miller. 1982. *Love Canal: The Social Construction of Disaster*. Washington, DC: Federal Emergency Management Agency.

Freeman, Harry. 1987. *Innovative Thermal Processes for Treating Hazardous Waste*. Lancaster, PA: Technomic.

Freudenberg, Nicholas. 1984. *Not in Our Backyards! Community Action for Health and the Environment*. New York: Monthly Review Press.

Freudenburg, William R., and Rodney K. Baxter. 1983. "Public Attitudes toward Local Nuclear Power Plants: A Reassessment." Paper presented at the annual meeting of the American Sociological Association.

Friedland, Steven I. 1981. "The New Hazardous Waste Managment System: Regulation of Wastes or Wasted Regulation?" *Harvard Environmental Law Review* 5:89-129.

Galida, Gary R. 1984. "Pennsylvania's Control Program for Hazardous Waste," pp. 242-253 in Majumdar and Miller, 1984.

Gallagher, Dorothy. 1979. "The Tragedy of Love Canal." *Redbook*, April, pp. 42, 67-71.

Gallup, George H. 1972. *The Gallup Poll: Public Opinion 1935-1971*. New York: Random House.

_____. 1978. *The Gallup Poll: Public Opinion 1972-1977*. Wilmington, DE: Scholarly Resources.

_____. 1979. *The Gallup Poll: Public Opinion 1978*. Wilmington, DE: Scholarly Resources.

_____. 1980. *The Gallup Poll: Public Opinion 1979*. Wilmington, DE: Scholarly Resources.

_____. 1981. *The Gallup Poll: Public Opinion 1980*. Wilmington, DE: Scholarly Resources.

Gamson, William A., and Andre Modigliani. 1989. "Media Discourse and Public Opinion on Nuclear Power: A Constructionist Approach." *American Journal of Sociology* 95(1):1-37.

Gans, Herbert J. 1979. *Deciding What's News: A Study of CBS Evening News, NBC Nightly News, Newsweek and Time*. New York: Pantheon.

Geiser, Ken, and Gerry Waneck. 1983. "PCBs and Warren County." *Science for the People* 15(4):13-17.

Gibbs, Lois Marie. 1982. *Love Canal: My Story*. Albany: State University of New York Press.

Gille, Zsuzsanna. 1992. "The Invisible Back-Scratching Hand of State Socialism: The Pattern of Nature Transformation and the Construction of the Environmental Problem in Former Socialist Countries." Unpublished M.A. thesis, University of California, Santa Cruz.

Gladwin, Thomas N. 1987. "Patterns of Environmental Conflict over Industrial Facilities in the United States, 1970-78," pp. 14-44 in Lake, 1987.

Gold, Allan R. 1989. "States' Deadline Today on Toxic-Waste Plans." *New York Times*, October 17.

Goldman, Benjamin A., James A. Hulme, and Cameron Johnson. 1986. *Hazardous Waste Management: Reducing the Risk*. Washington, DC: Island.

Goldman, L. R., B. Paigen, M. M. Magnant, and J. H. Highland. 1985. "Low Birth Weight, Prematurity and Birth Defects in Children Living Near the Hazardous Waste Site, Love Canal." *Hazardous Waste and Hazardous Materials* 2(2): 209-224.

Governor's Commission on Science and Technology for the State of New Jersey. 1983. *Report of the Governor's Commission on Science and Technology*. Trenton, NJ: Author.

Greenpeace. *Greenpeace*, various issues.

_____. 1988. *Greenpeace Toxics: Hazardous Waste Incinerators*. Washington, DC: Author.

Greer, Linda, et al. 1988. *Right Turn, Wrong Track: Failed Leadership in the Superfund Cleanup Program*. New York: Hazardous Waste Treatment Council.

Grisham, Joe W., ed. 1986. *Health Aspects of the Disposal of Waste Chemicals*. New York: Pergamon.

Grozuczak, Joanne. 1982. *Poisons on the Job: The Reagan Administration and American Workers* (Natural Heritage Report No. 4). San Francisco: Sierra Club.

Gwynne, Peter, Mark Whitaker, Elaine Shannon, Mary Hager, and Sharon Begley. 1978. "The Chemicals around Us." *Newsweek*, August 21, pp. 25-28.

Habermas, Jürgen. 1975. *Legitimation Crisis*. Boston: Beacon.

_____. 1984. *Theory of Communicative Action* (vol. 1). Boston: Beacon.

Hadden, Susan G., Joan Veillette, and Thomas Brandt. 1983. "State Roles in Siting Hazardous Waste Disposal Facilities: From State Preemption to Local Veto," pp. 196-211 in Lester and Bowman, 1983.

Hamilton, Cynthia. 1990. "Women, Home and Community: The Struggle in an Urban Environment." *Race, Poverty and the Environment* 1(1):3, 10-13.

Hanneman, Richard L. 1982. "A Service Industry Perspective," pp. 1-12 in Sweeney, Bhatt, Sykes, and Sproul, 1982.

Harris, Christopher, William L. Want, and Morris A. Ward. 1987. *Hazardous Waste: Confronting the Challenge*. New York: Quorum.

Harris, Robert H., Joseph H. Highland, Joseph V. Rodericks, and Stavros S. Papadopulos. 1984. "Adverse Health Effects at a Tennessee Hazardous Waste Disposal Site." *Hazardous Waste and Hazardous Materials* 1(2):183-204.

Harthill, Michalann, ed. 1984. *Hazardous Waste Management: In Whose Backyard?* Boulder, CO: Westview.

Harvey, David. 1989. *The Condition of Postmodernity*. Cambridge, MA: Basil Blackwell.

Hazardous Waste Dialogue Group, Program for Environmental Dispute Resolution, the Conservation Foundation. 1983. *Siting Hazardous Waste Management Facilities: A Handbook*. Washington, DC: Conservation Foundation.

Hedeman, William N., Jr., Paul E. Shorb III, and C. A. McLean. 1987. "The Superfund Amendments and Reauthorization Act of 1986: Statutory Provisions and EPA Implementation." *Hazardous Waste and Hazardous Materials* 4(2):193-210.

Heimer, Carol A. 1988. "Social Structure, Psychology, and the Estimation of Risk," pp. 491-519 in W. Richard Scott and Judith Blake, eds., *Annual Review of Sociology* (vol. 14). Palo Alto, CA: Annual Reviews.

Horkheimer, Max, and Theodor Adorno. 1987. *The Dialectics of Enlightenment*. New York: Continuum.

Huang, C. P., ed. 1981. *Industrial Waste: Proceedings of the Thirteenth Mid-Atlantic Conference*. Ann Arbor, MI: Ann Arbor Science Publishers.

Huggins, R. G., and A. D. Towery. 1987. "Waste Minimization: A Case Study." *Hazardous Waste and Hazardous Materials* 4(1):43-45.

Iyengar, Shanto, and Donald R. Kinder. 1987. *News That Matters: Television and American Opinion*. Chicago: University of Chicago Press.

Jameson, Frederic. 1984. "Postmodernism, or the Cultural Logic of Late Capitalism." *New Left Review*, 146.

_____. 1991. *Postmodernism, or the Cultural Logic of Late Capitalism*. Durham, NC: Duke University Press.

Johnnie, Susan T. 1987. "Waste Reduction in the Hewlett-Packard, Colorado Springs Division, Printed Circuit Board Manufacturing Shop." *Hazardous Waste and Hazardous Materials* 4(1):9-22.

Jones, Alex S. 1990. "Survey Finds That Americans Want News but Are Not Very Well Informed." *New York Times*, July 15.

Kaplan, E. Ann, ed. 1988. *Postmodernism and Its Discontents: Theories, Practices*. London: Verso.

Keystone Center. 1980. *Siting Non-radioactive Hazardous Waste Facilities: An Overview. Final Report of the First Keystone Workshop on Managing Non-radioactive Hazardous Wastes*. Keystone, CO: Author.

_____. 1984. *The Keystone Siting Process Handbook: A New Approach to Siting Hazardous Waste Management Facilities* (Report LP-194). Austin: Texas Dept. of Water Resources.

Kolata, Gina B. 1980. "Love Canal: False Alarm Caused by Botched Study." *Science* 208:1239-1242.

Kolbert, Elizabeth. 1992. "Over 'Murphy Brown,' Art Is Bigger Than Life," *New York Times*, September 23.

Kovacs, William L., and John F. Klucsik. 1977. "The New Federal Role in Solid Waste Management: The Resource Conservation and Recovery Act of 1976." *Columbia Journal of Environmental Law* 3:205-261.

Kraft, Michael E., and Ruth Kraut. 1988. "Citizen Participation and Hazardous Waste Policy Implementation," pp. 63-80 in Davis and Lester, 1988.

Krag, Bruce L. 1985. "Hazardous Wastes and Their Management." *Hazardous Waste and Hazardous Materials* 2(3):251-308.

Kroll-Smith, Steve, and Stephen R. Couch. 1991. "Social Impacts of Toxic Contamination." Report prepared for the Nevada Agency for Nuclear Projects, Yucca Mountain Socioeconomic Project, September.

Lake, Robert W., ed. 1987. *Resolving Locational Conflict*. New Brunswick, NJ: Center for Urban Policy Research.

Lane, Robert. 1966. *The Regulation of Businessmen: Social Conditions of Government Economic Control*. Hamden, CT: Archon.

Lash, Jonathan, Katherine Gillman, and David Sheridan. 1984. *A Season of Spoils: The Story of the Reagan Administration's Attack on the Environment*. New York: Pantheon.

Lash, Scott. 1990. *Sociology of Postmodernity*. London: Routledge.

Lave, Lester B. 1981. *The Strategy of Social Regulation: Decision Frameworks for Policy*. Washington, DC: Brookings Institution.

League of Women Voters. 1980. *A Hazardous Waste Primer* (LWV 545). Washington, DC: LWV Education Fund.

Legislative Commission on Toxic Substances and Hazardous Wastes (New York). 1987. *Hazardous Waste Facility Siting: A National Survey*. Albany: Author.

Lehman, John P. 1984. "Resource Conservation and Recovery Act of 1976," pp. 7-26 in Harthill, 1984.

Lennett, David J., and Linda E. Greer. 1985. "State Regulation of Hazardous Waste." *Ecology Law Quarterly* 12(2):183-269.

Lester, James P. 1983. "The Process of Hazardous Waste Regulation: Severity, Complexity, and Uncertainty," pp. 3-23 in Lester and Bowman, 1983.

Lester, James P., and Ann O. Bowman, eds. 1983. *The Politics of Hazardous Waste Management*. Durham, NC: Duke University Press.

Levine, Adeline G. 1982. *Love Canal: Science, Politics, and People*. Lexington, MA: Lexington.

Lewis, Sanford, and Marco Kaltofen. 1989. *From Poison to Prevention: A White Paper on Replacing Hazardous Waste Facility Siting with Toxics Reduction*. Boston: National Toxics Campaign Fund.

Lewis, Victor. 1990. "A Challenge to the Environmental Movement." *Race, Poverty and the Environment* 1(1):4, 19.

Lieber, Harvey. 1983. "Federalism and Hazardous Waste Policy," pp. 60-72 in Lester and Bowman, 1983.

Lilley, William, III, and James C. Miller III. 1977. "The New 'Social Regulation.' " *The Public Interest* 47(Spring):49-61.

Lindblom, C. 1959. "The Science of 'Muddling Through.' " *Public Administration Review* 19:79-88.

Lowrance, William W., ed. 1981. *Assessment of Health Effects at Chemical Disposal Sites: Proceedings of a Symposium Held in New York City on June 1-2, 1981 by the Life Sciences and Public Policy Program of the Rockefeller University.* Los Altos, CA: William Kaufmann.

MacAvoy, Paul W. 1979. *The Regulated Industries and the Economy.* New York: W. W. Norton.

Macbeth, Angus. n.d. "Private Industry: Compliance and Liability under the Environmental Laws." Unpublished paper, Sidley & Austin, Washington, D.C.

Magnuson, Ed, Peter Stoler, and J. Madeleine Nash. 1980. "The Poisoning of America: Belatedly, the Campaign Begins to Control Hazardous Chemical Wastes." *Time*, September 22, pp. 58-69.

Majumdar, Shyamal K., and E. Willard Miller, eds. 1984. *Hazardous and Toxic Wastes: Technology, Management and Health Effects.* Easton: Pennsylvania Academy of Science.

Marcuse, Herbert. 1964. *One-Dimensional Man: Studies in the Ideology of Advanced Industrial Society.* Boston: Beacon.

Martin, Donald L., and Warren F. Schwartz. 1977. *Deregulating American Industry.* Lexington, MA: D. C. Heath.

Marx, Karl. 1867/1967. *Capital: A Critique of Political Economy* (vol. 1). New York: International.

Matheny, Albert R., and Bruce A. Williams. 1988. "Rethinking Participation: Assessing Florida's Strategy for Siting Hazardous Waste Disposal Facilities," pp. 37-52 in Davis and Lester, 1988.

Maugh, Thomas H., II. 1982. "Just How Hazardous Are Dumps?" *Science* 215:490-493.

Mazur, Allan. 1981. *The Dynamics of Technical Controversy.* Washington, DC: Communications Press.

_____. 1984a. "The Journalists and Technology: Reporting about Love Canal and Three Mile Island." *Minerva* 22(1):45-66.

_____. 1984b. "Media Influences on Public Attitudes toward Nuclear Power," pp. 97-114 in William R. Freudenburg and Eugene A. Rosa, eds., *Public Reactions to Nuclear Power: Are There Critical Masses?* Boulder, CO: Westview.

McAdam, Doug, John D. McCarthy, and Mayer N. Zald. 1988. "Social Movements," pp. 695-737 in Neil J. Smelser, ed., *The Handbook of Sociology.* Newbury Park, CA: Sage.

McAvoy, J. F. 1980. "Hazardous Waste Management in Ohio: The Problem of Siting." *Capital University Law Review* 9:435-450.

McCoy and Associates, Inc. 1985. "The Outlook for Commercial Hazardous Waste Management Facilities: A Nationwide Perspective." *Hazardous Waste Consultant*, March/April, pp. 4-1–4-4.

_____. 1986. "1986 Outlook for Commercial Hazardous Waste Management Facilities: A Nationwide Perspective." *Hazardous Waste Consultant*, March/April, pp. 4-1–4-6.

_____. 1987. "1987 Outlook for Commercial Hazardous Waste Management Facilities: A Nationwide Perspective." *Hazardous Waste Consultant*, March/April, pp. 4-1–4-5.

_____. 1988. "1988 Outlook for Hazardous Waste Management Facilities: A Nationwide Perspective." *Hazardous Waste Consultant*, March/April, pp. 4-1–4-5.

_____. 1989. "1989 Outlook for Commercial Hazardous Waste Management Facilities: A Nationwide Perspective." *Hazardous Waste Consultant*, March/April, pp. 4-1–4-6.

_____. 1990. "1990 Outlook for Commercial Hazardous Waste Management Facilities: A Nationwide Perspective." *Hazardous Waste Consultant*, March/April, pp. 4-1–4-6.

McGuire, James E. 1986. "The Dilemma of Public Participation in Facility Siting Decisions and the Mediation Alternative." *Seton Hall Legislative Journal* 9(2):467-473.

Melloan, George, and Joan Melloan. 1978. *The Carter Economy*. New York: Wiley.

Miliband, Ralph. 1969. *The State in Capitalist Society*. New York: Basic Books.

Mitchell, Robert Cameron. 1984a. "Public Opinion and Environmental Politics in the 1970s and 1980s," pp. 51-74 in Norman J. Vig and Michael E. Kraft, eds., *Environmental Policy in the 1980s: Reagan's New Agenda*. Washington, DC: Congressional Quarterly Press.

_____. 1984b. "Rationality and Irrationality in the Public's Perception of Nuclear Power," pp. 137-179 in William R. Freudenburg and Eugene A. Rosa, eds., *Public Reactions to Nuclear Power: Are There Critical Masses?* Boulder, CO: Westview.

Mitnick, Barry M. 1980a. *The Political Economy of Regulation*. New York: Columbia University Press.

_____. 1980b. "Myths of Creation and Fables of Administration: Explanation and the Strategic Use of Regulation." *Public Administration Review* 40(3):275-286.

Moloch, Harvey, and Marilyn Lester. 1975. "Accidental News: The Great Oil Spill as Local Occurrence and National Event." *American Journal of Sociology* 81(2):235-260.

Montague, Peter. 1989. "What We Must Do: A Grass-Roots Offensive against Toxics in the '90s." *The Workbook* 14(3):90-113.

Morganthau, Tom, and Mary Hager. 1980. "Coping with Toxic Waste." *Newsweek*, May 19, pp. 34-35.

Morrell, David L. 1987. "Siting and the Politics of Equity," pp. 117-136 in Lake, 1987.

Morrell, David L., and Christopher Magorian. 1982. *Siting Hazardous Waste Facilities: Local Opposition and the Myth of Preemption*. Cambridge, MA: Ballinger.

Mugdan, Walter E., and Bruce R. Adler. 1985. "The 1984 RCRA Amendments: Congress as a Regulatory Agency." *Columbia Journal of Environmental Law* 10(2):215-254.

Nader, Ralph. 1965. *Unsafe at Any Speed: The Designed-In Dangers of the American Automobile*. New York: Pocket Books.

Nader, Ralph, R. Brownstein, and J. Richards. 1981. *Who's Poisoning America: Corporate Polluters and Their Victims in the Chemical Age*. San Francisco: Sierra Club Books.

National Campaign Against Toxic Hazards. 1988. *The Citizens Toxics Protection Manual*. Boston: Author.

National Conference of State Legislatures. 1982. *Hazardous Waste Management: A Survey of State Legislation 1982*. Denver: Author.

National Toxics Campaign. n.d. *Toxics Prevention: A Citizens' Platform*. Boston: Author.

_____. 1989. *Toxics Use Reduction: From Pollution Control to Pollution Prevention*. Boston: PIRG Toxics Action and the National Toxics Campaign.

_____. *Toxic Times*, various issues.

National Wildlife Federation. 1982. *Shredding the Environmental Safety Net: The Full Story behind the EPA Budget Cuts*. Washington, DC: Author.

_____. 1986. *Annual Report*. Washington, DC: Author.

Neumann, W. Russell. 1986. *The Paradox of Mass Politics: Knowledge and Opinion in the American Electorate*. Cambridge, MA: Harvard University Press.

Nugent, Angela. 1987. "The Power to Define a New Disease: Epidemiological Politics and Radium Poisoning," pp. 177-191 in Rosner and Markowitz, 1987b.

O'Connor, James. 1973. *The Fiscal Crisis of the State*. New York: St. Martin's.

Offe, Claus. 1974. "Structural Problems of the Capitalist State." *German Political Studies* 1:31-57.

_____. 1975. "The Theory of the Capitalist State and the Problem of Policy Formation," pp. 125-144 in Leon N. Lindberg, et al., eds., *Stress and Contradiction in Modern Capitalism: Public Policy and the Theory of the State*. Lexington, MA: Lexington.

O'Hare, Michael. 1977. " 'Not on *My* Block You Don't': Facilities Siting and the Strategic Importance of Compensation." *Public Policy* 25(4):407-458.

O'Hare, Michael, Lawrence Bacow, and Debra Sanderson. 1983. *Facility Siting and Public Opposition.* New York: Van Nostrand-Reinhold.

Oldenburg, Kirsten U., and Joel S. Hirschhorn. 1987. "Waste Reduction: From Policy to Commitment." *Hazardous Waste and Hazardous Materials* 4(1):1-8.

Oreskes, Michael. 1984. "Witness Says Crime Figures Rule Disposal of Toxic Waste." *New York Times*, September 20.

_____. 1990a. "America's Politics Loses Way as Its Vision Changes World." *New York Times*, March 18.

_____. 1990b. " 'Wars' Wound Candidates and the Process." *New York Times*, March 19.

_____. 1990c. "Study Finds 'Astonishing' Indifference to Elections." *New York Times*, May 6.

Ottinger, Richard. 1985. "Strengthening of the Resource Conservation and Recovery Act in 1984: The Original Loopholes, the Amendments, and the Political Factors Behind Their Passage." *Pace Environmental Law Review* 3(1):1-28.

Owen, Bruce M., and Ronald Braeutigam. 1978. *The Regulation Game: Strategic Use of the Administrative Process.* Cambridge, MA: Ballinger.

Paigen, Beverly, Goldman, L. R., J. H. Highland, M. M. Magnant, and A. T. Steegman, Jr. 1985. "Prevalence of Health Problems in Children Living near Love Canal." *Hazardous Waste and Hazardous Materials* 2(1): 23-44.

People Against Hazardous Landfill Sites. *PAHLS* (quarterly newsletter), various issues.

Perelman, Michael. 1990. "Marx as a Natural Resource Theorist." Paper presented at the Working Conference on Environment and Society, Santa Cruz, CA.

Pertschuk, Michael. 1982. *Revolt against Regulation: The Rise and Pause of the Consumer Movement.* Berkeley: University of California Press.

Petros, James K., Jr. 1987. "Waste Minimization Efforts at Union Carbide Corporation." *Hazardous Waste and Hazardous Materials* 4(1):47-53.

Piasecki, Bruce W. and Gary A. Davis. 1987. *America's Future in Toxic Waste Management: Lessons from Europe.* New York: Quorum.

Pirages, Suellen. 1987. "HW: Anywhere but Here." *State Legislatures* 13(4):34.

Polanyi, Karl. 1944. *The Great Transformation: The Political and Economic Origins of Our Time.* Boston: Beacon.

Popper, Frank J. 1987a. "The Environmentalists and the LULU," pp. 1-13 in Lake, 1987.

_____. 1987b. "LP/HC and LULUs: The Political Uses of Risk Analysis in Land-Use Planning," pp. 275-287 in Lake, 1987.

Portney, Kent E. 1988. "The Role of Economic Factors in Lay Perceptions of Risk," pp. 53-62 in Davis and Lester, 1988.

Poulantzas, Nicos. 1975. *Political Power and Social Class.* London: New Left.

Purcell, Arthur, ed. 1986. *Hazardous and Solid Waste Minimization.* Washington, DC: Government Institutes.

Rankin, William L., Stanley M. Nealey, and Barbara Desow Melber. 1984. "Overview of National Attitudes toward Nuclear Energy: A Longitudinal Analysis," pp. 41-68 in William R. Freudenburg and Eugene A. Rosa, eds. *Public Reactions to Nuclear Power:. Are There Critical Masses?* Boulder, CO: Westview.

Robbins, Richard L. 1982. "Methods to Gain Community Support for a Hazardous Waste Facility or a Superfund Cleanup," pp. 503-518 in Sweeney, Bhatt, Sykes and Sproul, 1982.

Rosner, David, and Gerald Markowitz. 1987a. " 'A Gift of God'? The Public Health Controversy over Leaded Gasoline during the 1920s," pp. 121-139 in Rosner and Markowitz, 1987b.

_____. 1987b. *Dying for Work: Workers' Safety and Health in Twentieth-Century America.* Bloomington: Indiana University Press.

Ross, Andrew, ed. 1988. *Universal Abandon? The Politics of Postmodernism.* Minneapolis: University of Minnesota Press.

Ruckelshaus, William D. 1984a. "E.P.A. Battle on Hazardous Waste Is in Full Swing." Letter to the *New York Times*, May 21.

——. 1984b. "The Role of the Affected Community in Superfund Cleanup Activities." *Hazardous Waste and Hazardous Materials* 1(3):283-288.

Ruffins, Paul. 1990. "Blacks and Greens: What Can the Environmental Movement Do to Reach Out to Minorities?" *Race, Poverty and the Environment* 1(2):5.

Rutter, Nancy. 1991. "The Greening of Corporate America." *California Lawyer* 2(4): 33-36.

Ryan, Ann Sprightly. 1984. "Approaches to Hazardous Waste Facility Siting in the United States." Report to the Massachusetts Hazardous Waste Facility Site Safety Council, Boston.

Sandman, Peter M. 1987. "Getting to Maybe: Some Communications Aspects of Siting Hazardous Waste Facilities," pp. 324-344 in Lake, 1987.

Sarokin, David. 1987. "Going to the Source: The Real Solution to the Toxic Waste Crisis." *Greenpeace* 12(1):16-18.

Schneider, Joseph W. 1985. "Social Problems Theory: The Constructionist View," pp. 209-211 in Ralph H. Turner et al., eds., *Annual Review of Sociology* (vol. 11). Palo Alto: Annual Reviews.

Schoenfeld, A. Clay, Rober F. Meier, and Robert J. Griffin. 1979. "Constructing a Social Problem: The Press and the Environment." *Social Problems* 27:38-61.

Sewell, W. P. Derrick, and Timothy O'Riordan. 1976. "The Culture of Participation in Environmental Decisionmaking." *Natural Resources Journal* 16(1):1-21.

Shabecoff, Philip. 1979. "House Unit Attacks Lags on Toxic Waste." *New York Times*, October 14.

——. 1983. "E.P.A. Says It Maps New Top Priority." *New York Times*, November 21.

——. 1989. "Where the Planet Is Losing Its Life Forms." *New York Times*, July 30.

——. 1990a. "40-Year Countdown Is Seen to Rescue the Environment." *New York Times*, February 11.

——. 1990b. "E.P.A. Finds Much of Love Canal Safe." *New York Times*, May 15.

Sheilds, Merrill, William J. Cook, Mary Hager, and John Carey. 1980. "Fiery Cloud over Earth Day." *Newsweek*, May 5, pp. 80-82.

Sheoin, Tomas Mac. 1985. "The Dismantling of US Health and Safety Regulations under the First Reagan Administration: A Bibliography." *International Journal of Health Services* 15(4):585-608.

Short, James. 1984. "The Social Fabric at Risk: Toward the Social Transformation of Risk Analysis." *American Sociological Review* 49(6):711-725.

Sierra Club. 1987. *Implementation of Toxic Control Laws*. San Francisco: Author.

Silicon Valley Toxics Coalition. *Silicon Valley Toxics News* (newsletter), various issues.

Simon, Philip J. 1983. "Reagan in the Workplace: Unraveling the Health and Safety Net." Report by the Center for Study of Responsive Law, Washington, DC.

Slovic, Paul, Baruch Fischhoff, and Sarah Lichtenstein. 1980. "Facts versus Fears: Understanding Perceived Risk," pp. 181-214 in Richard C. Schwing and Walter A. Albers, Jr., eds., *Societal Risk Assessment: How Safe Is Safe Enough?* New York: Plenum.

——. 1985. "Characterizing Perceived Risk," pp. 91-125 in Robert W. Kates, Christoph Hohenemser, and Jeanne X. Kasperson, eds., *Perilous Progress: Managing the Hazards of Technology*. Boulder, CO: Westview.

Smith, Martin A., Frances M. Lynn, Richard N. L. Andrews, Richard Olin, and Cassandra Maurer. 1985. *Costs and Benefits to Local Government Due to Presence of a Hazardous Waste Management Facility and Related Compensation Issues*. Chapel Hill: Institute for Environmental Studies, University of North Carolina.

Smith, Paul. 1988. "Visiting the Banana Republic," pp. 128-148 in Ross, 1988.

Smothers, Ronald. 1989. "50 Waste Disposal Plans Help and Hinder Project." *New York Times*, October 27.

_____. 1991. "Future in Mind, Choctaws Reject Plan for Landfill." *New York Times*, April 21.

Southern Organizing Committee for Economic and Social Justice. 1992. "Call for a Community/Labor Conference for Environmental Justice."

Spector, Malcolm, and John I. Kitsuse. 1987. *Constructing Social Problems*. New York: Aldine deGruyter.

Starr, Chauncey. 1969. "Social Benefit versus Technological Risk." *Science* 165(September): 1232-1238.

_____. 1980. "The Risks We Run and the Risks We 'Accept,' " pp. 1-4 in Richard C. Schwing and Walter A. Albers, Jr., eds., *Societal Risk Assessment: How Safe Is Safe Enough?* New York: Plenum.

_____. 1985. "Risk Management, Assessment and Acceptability." *Risk Analysis* 5(2):97-102.

Stern, Paul C., Thomas Dietz, and J. Stanley Black. 1986. "Support for Environmental Protection: The Role of Moral Norms." *Population and Environment* 8(3/4):204-222.

Stevens, William K. 1990. "If It's East of the Mississippi, It's Blanketed in Pollution's Haze." *New York Times*, July 17.

Stigler, George J. 1975. *The Citizen and the State*. Chicago: University of Chicago Press.

Stockman, David, and Jack Kemp. 1980. "Avoiding an Economic Dunkirk." *New York Times*, December 14.

Stone, Peter H. 1981. "Conservative Brain Trust." *New York Times Magazine*, May 5, pp. 18-93.

Sturrock, John. 1986. *Structuralism*. London: Paladin Grafton.

Susskind, Lawrence E. 1985. "The Siting Puzzle: Balancing Economic and Environmental Gains and Losses." *Environmental Impact Assessment Review* 5(2):157-163.

Sweeney, Thomas L., Harasiddhiprasad G. Bhatt, Robert M. Sykes, and Otis J. Sproul, eds. 1982. *Hazardous Waste Management for the 80s*. Ann Arbor, MI: Ann Arbor Science Publishers.

Szasz, Andrew. 1982. "The Dynamics of Social Regulation: A Study of the Formation and Evolution of the Occupational Safety and Health Administration." Unpublished doctoral dissertation, University of Wisconsin—Madison.

_____. 1984. "Industrial Resistance to Occupational Safety and Health Legislation: 1971-1981." *Social Problems* 32(2):103-116.

_____. 1986a. "The Process and Significance of Political Scandals: A Comparison of Watergate and the 'Sewergate' Episode at the Environmental Protection Agency." *Social Problems* 33(3):202-217.

_____. 1986b. "Corporations, Organized Crime, and the Disposal of Hazardous Waste: An Examination of the Making of a Criminogenic Regulatory Structure." *Criminology* 24(1):1-27.

_____. 1986c. "The Reversal of Federal Policy toward Worker Safety and Health: A Critical Examination of Alternative Explanations." *Science and Society* 50(1):25-51.

_____. 1988. "Risk Perception and Modern Social Theory." Paper presented at the annual meeting of the American Sociological Association, Atlanta, GA.

_____. 1992. "In Praise of Policy Luddism: Strategic Lessons from the Hazardous Waste Wars." *Capitalism, Nature, Socialism: A Journal of Socialist Ecology* 2(1):17-43.

Tarlock, A. Dan. 1984. "Siting New or Expanded Treatment, Storage or Disposal Facilities: The Pigs in the Parlors of the 1980s." *Natural Resources Lawyer* 17:429-461.

Tarlock, A. Dan, and Robert L. Glicksman. 1987. "Hazardous Waste Sites: New Opportunities for Local Control under Superfund." *Zoning and Planning Law Report* 10(7):137-142.

Taylor, Stuart, Jr. 1983. "E.P.A. Inquiries Center on Four Issues." *New York Times*, March 13.

Texans United. 1989. *Texas Report* (newsletter), June.

Thomas, Ginny, and Bill Brooks. 1981. " 'Buddy, We're Home': Halting the Heard Landfill." *Southern Exposure* 9(3):38-41.

Thomas, Kenny. 1981. " 'Our Government Wouldn't Do This to Us.' " *Southern Exposure* 9(3):28-33.

Thompson, E. P. 1963. *The Making of the English Working Class*. New York: Vintage.

Toxics Coordinating Project, California Toxics Coalition. n.d. *Toxics Use Reduction: Reduce It, Don't Produce It*. San Francisco: California Toxics Coalition.

Tucker, Robert C. 1978. *The Marx-Engels Reader* (2nd ed.). New York: W. W. Norton.

Tversky, Amos, and Daniel Kahneman. 1982. "Judgement under Uncertainty: Heuristics and Biases," pp. 3-20 in Daniel Kahneman, Paul Slovic, and Amos Tversky, eds., *Judgement under Uncertainty: Heuristics and Biases*. Cambridge: Cambridge University Press.

U.S. Bureau of the Census. 1980. *Statistical Abstracts of the United States, 1980*. Washington, DC: Government Printing Office.

U.S. Congress, Office of Technology Assessment. 1986. *Serious Reduction of Hazardous Waste: Summary*. Washington, DC: Government Printing Office.

_____. 1987. *From Pollution to Prevention: A Progress Report on Waste Reduction—Special Report* (OTA-ITE-347). Washington, DC: Government Printing Office.

_____. 1988. *Are We Cleaning Up?* Washington, DC: Government Printing Office.

_____. 1989. *Cleaning Up: Superfund's Problems Can Be Solved*. Washington, DC: Government Printing Office.

U.S. Council on Environmental Quality. 1980. *Public Opinion on Environmental Issues: Results of a National Public Opinion Survey*. Washington, DC: Government Printing Office.

U.S. Environmental Protection Agency. 1973. *Public Attitudes toward Hazardous Waste Disposal Facilities*. Washington, DC: Government Printing Office.

_____. 1974. *Disposal of Hazardous Wastes* (Report to Congress pursuant to Section 212 of the Solid Waste Disposal Act, as amended). Washington, DC: Government Printing Office.

_____. 1976. *Hazardous Waste Management: Public Meetings*, December 2-11. Washington, DC: Government Printing Office.

_____. 1979. *Siting of Hazardous Waste Management Facilities and Public Opposition* (SW-809). Washington, DC: Government Printing Office.

_____. 1980. *Hazardous Waste Facility Siting: A Critical Problem* (SW-865). Washington, DC: Government Printing Office.

_____. 1982a. *Using Compensation and Incentives When Siting Hazardous Waste Management Facilities* (SW-942). Washington, DC: Government Printing Office.

_____. 1982b. *Using Mediation When Siting Hazardous Waste Management Facilities* (SW-944). Washington, DC: Government Printing Office.

_____. 1986. *Report to Congress: Minimization of Hazardous Waste*. Washington, DC: Government Printing Office.

_____. 1991. *Toxics in the Community: National and Local Perspectives. The 1989 Toxics Release Inventory National Report* (560/4-91-014). Washington, DC: Government Printing Office.

U.S. General Accounting Office. 1978a. *How to Dispose of Hazardous Waste: A Serious Question That Needs to Be Resolved* (CED-79-13). Washington, DC: Government Printing Office.

_____. 1978b. *Waste Disposal Practices: A Threat to Health and the Nation's Water Supply* (CED-78-120). Washington, DC: Government Printing Office.

_____. 1979. *Hazardous Waste Management Programs Will Not Be Effective: Greater Efforts Are Needed* (CED-79-14). Washington, DC: Government Printing Office.

_____. 1981. *Hazardous Waste Facilities with Interim Status May Be Endangering Public Health and the Environment* (CED-81-158). Washington, DC: Government Printing Office.

_____. 1982. *Environmental Protection Agency's Progress Implementing the Superfund Program* (CED 82-91). Washington, DC: Government Printing Office.

_____. 1983a. *Evaluation of the Environmental Protection Agency's Inspector General Audit of Superfund Expenditures and Implementation of the Inspector General's Recommendations* (RCED-84-31) Washington, DC: Government Printing Office.

_____. 1983b. *Siting of Hazardous Waste Landfills and Their Correlation with Racial and Economic Status of Surrounding Communities* (GAO/RCED-83-168). Washington, DC: Government Printing Office.

_____. 1983c. *Interim Report on Inspection, Enforcement, and Permitting Activities at Hazardous Waste Facilities* (RCED-83-241). Washington, DC: Government Printing Office.

_____. 1985a. *EPA's Inventory of Potential Hazardous Waste Sites Is Incomplete* (RCED-85-75). Washington, DC: Government Printing Office.

_____. 1985b. *Hazardous Waste: Status of Private Party Efforts to Clean Up Hazardous Waste Sites* (RCED-86-65FS). Washington, DC: Government Printing Office.

_____. 1985c. *Illegal Disposal of Hazardous Waste: Difficult to Detect or Deter* (RCED-85-2). Washington, DC: Government Printing Office.

_____. 1985d. *Status of EPA's Remedial Cleanup Efforts* (RCED-85-86). Washington, DC: Government Printing Office.

_____. 1986a. *Hazardous Waste: Environmental Safeguards Jeopardized When Facilities Cease Operation* (RCED-86-77). Washington, DC: Government Printing Office.

_____. 1986b. *Hazardous Waste: EPA Has Made Limited Progress in Determining the Wastes to Be Regulated* (RCED-87-27). Washington, DC: Government Printing Office.

_____. 1987a. *Hazardous Waste: Corrective Action Cleanups Will Take Years to Complete* (RCED-88-48). Washington, DC: Government Printing Office.

_____. 1987b. *Hazardous Waste: Facility Inspections Are Not Thorough and Complete* (RCED-88-20). Washington, DC: Government Printing Office.

_____. 1987c. *Hazardous Waste: Uncertainties of Existing Data* (PEMD-87-11BR). Washington, DC: Government Printing Office.

_____. 1988a. *Hazardous Waste: Groundwater Conditions at Many Land Disposal Facilities Remain Uncertain* (RCED-88-29). Washington, DC: Government Printing Office.

_____. 1988b. *Hazardous Waste: Many Enforcement Actions Do Not Meet EPA Standards* (RCED-88-140). Washington, DC: Government Printing Office.

_____. 1988c. *Hazardous Waste: New Approaches Needed to Manage the Resource Conservation and Recovery Act* (RCED-88-115). Washington, DC: Government Printing Office.

U.S. House of Representatives. 1974a. *Amendments to the Solid Waste Disposal Act*. Prepared by the staff for the use of the Committee on Interstate and Foreign Commerce and its Subcommittee on Public Health and Environment, 93rd Congress, 2d Session, March. Committee Print 19.

_____. 1974b. *Solid Waste Disposal Act Extension—1974*. Hearings before the Subcommittee on Public Health and Environment, Committee on Interstate and Foreign Commerce, 93rd Congress, 2d Session, March 27-28. Serial 93-78.

_____. 1975. *Waste Control Act of 1975*. Hearings held by the Subcommittee on Transportation and Commerce, Committee on Interstate and Foreign Commerce, 94th Congress, 1st Session, April 8-11, 14-17. Serial 94-28.

_____. 1976. *Resource Conservation and Recovery Act of 1976*. House Report 94-1491 (Report on H.R. 14496). *U.S. Code Congressional and Administration News*, pp. 6238-6354.

_____. 1977. *Oversight—Resource Conservation and Recovery Act of 1976*. Subcommittee on Transportation and Commerce, Committee on Interstate and Foreign Commerce, 95th Congress, 1st Session, April 26, May 18-19. Serial 95-38.

_____. 1978a. *Resource Conservation and Recovery Act—Oversight*. Subcommittee on Transportation and Commerce, Committee on Interstate and Foreign Commerce, 95th Congress, 2d Session, March 7-9. Serial 95-124.

_____. 1978b. *Oversight—Resource Conservation and Recovery Act.* Hearings, Committee on Interstate and Foreign Commerce, Subcommittee on Oversight and Investigations, 95th Congress, 2d Session, October, 30. Serial 95-183.

_____. 1979a. *Hazardous Waste Disposal, Part 1.* Hearings, Committee on Interstate and Foreign Commerce, Subcommittee on Oversight and Investigations, 96th Congress, 1st Session, March 21-June 19. Serial 96-48.

_____. 1979b. *Hazardous Waste Disposal, Part 2.* Hearings, Committee on Interstate and Foreign Commerce, Subcommittee on Oversight and Investigations, 96th Congress, 1st Session, March 21-June 19. Serial 96-49.

_____. 1979c. *Hazardous Waste Disposal.* Subcommittee on Oversight and Investigations, Committee on Interstate and Foreign Commerce, 96th Congress, 1st Session, September. Committee Print 96-IFC-31.

_____. 1979d. *Resource Conservation and Recovery Act Authorization.* Hearings, Committee on Interstate and Foreign Commerce, Subcommittee on Transportation and Commerce, 96th Congress, 1st Session, March 27-28. Serial 96-31.

_____. 1980. *Organized Crime and Hazardous Waste Disposal.* Hearings, Subcommittee on Oversight and Investigations, Committee on Interstate and Foreign Commerce, 96th Congress, 2d Session, December 16.

_____. 1981a. *Organized Crime Links to the Waste Disposal Industry.* Hearings, Subcommittee on Oversight and Investigations, Committee on Energy and Commerce, 97th Congress, 1st Session, May 28.

_____. 1981b. *Hazardous Waste Matters: A Case Study of Landfill Sites.* Hearings, Subcommittee on Oversight and Investigations, Committee on Energy and Commerce, 97th Congress, 1st Session, June 9.

_____. 1982a. *Resource Conservation and Recovery Act Reauthorization.* Hearings, Subcommittee on Commerce, Transportation, and Tourism, Committee on Energy and Commerce, 97th Congress, 2d Session, March 31, April 21.

_____. 1982b. *Hazardous Waste Siting Problems: EPA Oversight.* Hearings, Subcommittee of the Committee on Government Operations, 97th Congress, 2d Session, April 23.

_____. 1983. *House Report (Energy and Commerce Committee) No. 98-198, Parts I and II* (to accompany H.R. 2867). 98th Congress, 1st Session, May 17, June 9.

_____. 1985. *Investigation of the Role of the Department of Justice in the Withholding of Environmental Protection Agency Documents from Congress in 1982-83* (4 vols.). Report of the Committee on the Judiciary, December 11.

_____. 1987. *Potential Hazardous Waste Volume and Capacity Problems.* Hearings, Subcommittee on Environment, Energy and Natural Resources, Committee on Government Operations, 99th Congress, 2d Session, September 24, 1986.

U.S. Senate. 1974. *The Need for a National Materials Policy.* Hearings, Subcommittee on Environmental Pollution, Committee on Public Works, 93rd Congress, 2d Session, June 11-13, July 9-11, 15-18. Serial 93-H47.

_____. 1976a. *Solid Waste Utilization Act of 1976.* Report of the Committee on Public Works. Senate Report 94-988 (on S. 3622).

_____. 1976b. Senate Floor Debate, 94th Congress, 2d Session, June 30. *Congressional Record* 122, Part 17, pp. 21393-21437.

_____. 1978. *Resource Conservation and Recovery Act Oversight.* Subcommittee on Resource Protection of the Committee on Environment and Public Works, 95th Congress, 2d Session, March 20. Serial 95-H57.

_____. 1979. *Oversight of Hazardous Waste Management and the Resource Conservation and Recovery Act.* Hearings, Subcommittee on Oversight of Governmental Management, Committee on Government Affairs, 96th Congress, 1st Session, July 19, August 1.

_____. 1980. *Report on Hazardous Waste Management and the Implementation of RCRA.* Hearings, Subcommittee on Oversight of Governmental Management, Committee on Government Affairs, 96th Congress, 2d Session, March.

_____. 1982. *Reauthorization of the Resource Conservation and Recovery Act.* Hearings, Subcommittee on Environmental Pollution of the Committee on Environment and Public Works, 97th Congress, 2d Session, April 26, June 24,

_____. 1983. *A Legislative History of the Comprehensive Environmental Response, Compensation, and Liability Act of 1980 (Superfund), Public Law 96-510* (3 vols.). Report of the Committee on Environment and Public Works.

_____. 1990. *A Legislative History of the Superfund Amendments and Reauthorization Act of 1986 (Public Law 99-499)* (7 vols.). Report of the Committee on Environment and Public Works. Senate Print 101-120.

Vanderbilt Television News Archive. *Television News Index and Abstracts* (various years). Nashville, TN: Author.

Vince, Peggy J. 1982. "The Hazardous Waste Management Triangle," pp. 17-25 in Sweeney, Bhatt, Sykes, and Sproul, 1982.

Walter, Benjamin ,and Malcolm Getz. 1986. "Social and Economic Effects of Toxic Waste Disposal," pp. 223-245 in Sheldon Kamieniecki, Robert O'Brien, and Michael Clarke, eds., *Controversies in Environmental Policy.* Albany: State University of New York Press.

Weidenbaum, Murray C. 1979. *The Future of Business Regulation.* New York: Amacon.

Weiss, Michael J. 1980. "Underground Time Bombs: Chemical Waste Dumps." *Mechanix Illustrated*, September, pp. 55, 96-101.

West, William L. 1985. "Hazardous Waste Management: An Industry Perspective," pp. 35-40 in Bhatt, Sykes, and Sweeney, 1985.

Wildavsky, Aaron. 1979. "No Risk Is the Highest Risk of All." *American Scientist* 67:32-37.

Windsor, Eleanor W. 1981. "Public Participation: The Missing Ingredient for Success in Hazardous Waste Siting," pp. 517-524 in Huang, 1981.

_____. 1983. "Siting Hazardous Waste Management Facilities: A New Approach Is Needed," pp. 1-7 in Michael D. LaGrega and Linda K. Hendrian, eds., *Toxic and Hazardous Wastes: Proceedings of the Fifteenth Mid-Atlantic Industrial Waste Conference.* Woburn, MA: Butterworth.

_____. 1987. "Public Relations and Participation: A Trail of Frustration with a Chance for Improvement," pp. 377-385 in Jeffrey C. Evans, ed., *Toxic and Hazardous Wastes: Proceedings of the Nineteenth Mid-Atlantic Industrial Waste Conference.* Lancaster, PA: Technomic.

Wolf, Sidney R. 1980. "Public Opposition to Hazardous Waste Sites: The Self-Defeating Approach to National Hazardous Waste Control under Subtitle C of the Resource Conservation and Recovery Act of 1976." *Boston College Environmental Affairs Law Review* 8(3):463-540.

Worobec, Mary. 1980. "An Analysis of the Resource Conservation and Recovery Act." *Environment Reporter* 11:633.

Wright, Erik O. 1979. *Class, Crisis and the State.* London: New Left.

Wurth-Hough, Sandra J. 1982. "Chemical Contamination and Governmental Policymaking: The North Carolina Experience." *State and Local Government Review* 14(2):54-60.

Zeff, Robbin L., and Sue Greer. 1988. *Making It Happen: Putting on a Leadership Development Conference* Arlington, VA: Citizen's Clearinghouse for Hazardous Wastes.

Zeff, Robbin L., Marsha Love, and Karen Stults. 1989. *Empowering Ourselves: Women and Toxics Organizing.* Arlington, VA: Citizen's Clearinghouse for Hazardous Wastes.

Legislation

Comprehensive Environmental Response, Compensation and Liability Act of 1980, P.L. 96-510.

Hazardous and Solid Waste Amendments of 1984, P.L. 98-616, 98 Stat. 3221.

Resource Conservation and Recovery Act of 1976, P.L. 94-580, 90 Stat. 2795-2841, in *U.S. Code Congressional and Administrative News*, 1976, vol. 2.

Resource Recovery Act of 1970, P.L. 91-512, 84 Stat. 1227, in *U.S. Code Congressional and Administrative News*, 91st Congress, 2d Session, 1970, pp. 1427-1437.
Solid Waste Disposal Act of 1965, P.L. 89-272, 79 Stat. 997.
Superfund Amendments and Reauthorization Act of 1986, P.L. 99-499, 100 Stat 1613.

Cases

City of Philadelphia v. New Jersey, 437 U.S. 617, 1978.
Hardage v. Atkins, 619 F.2d 871, 10th Circuit, 1980.
Industrial Union Department, AFL-CIO v. American Petroleum Institute, 448 U.S. 607, 1980.

Conferences

"Environmental Liability in Real Estate Transactions in California," by Professional Education Systems, Inc., April 24, 25, 26, 1990, San Francisco, San Jose, Sacramento, CA.
"Hazardous Waste in Real Estate Transactions," by CLE International, October 4-5, 1990, Los Angeles.
"Hazardous Waste Reduction," by the New York State Department of Environmental Conservation, at Albany, NY. Second Annual, June 13-14, 1989; Third Annual, June 19-20, 1990.

In-depth Interviews with Toxics Movement Organizers and Local Leaders

Note: All interviews were conducted by Hal Aronson.
Marty Chestnutt, Citizen's Clearinghouse for Hazardous Wastes, CCHW South organizer, September 10, 1990; October 26, 1992.
Will Collette, Citizen's Clearinghouse for Hazardous Wastes, national organizer, September 17, 1990.
Gary Cohen, National Toxics Campaign, September 28, 1989.
Luke Cole, California Rural Legal Assistance, September 18, 1990.
Lew Dunn, Greenpeace, California organizer, April 28, 1990.
Lois Gibbs, Citizen's Clearinghouse for Hazardous Wastes, October 30, 1992.
Sue Greer, People Against Hazardous Landfill Sites, September 21, 1989.
Kaye Kiker, cofounder of Alabamians for a Clean Environment and president of the National Toxics Campaign, October 8, 1990.
Sally Teets, Citizen's Clearinghouse for Hazardous Wastes, CCHW Midwest organizer, September 10, 1990.
John Thompson, Central States Resource Center, September 12, 1989.

Brief Interviews with Staff at National Headquarters of Environmental and Public Interest Organizations

Conservation Foundation, Lane Krahl, July 1989.
Environmental Defense Fund, Lois Epstein, July 1989.
Greenpeace, Margie Kelley, July 1989.
League of Women Voters, Lloyd Leonard, September 1989.
National Audubon Society, Ann Strickland, September 1989.
National Resources Defense Council, Don Strait, July 1989.
National Wildlife Federation, Gerry Poje, July 1989.
USPIRG, Rick Hind, July 1989.

A survey of state agencies was conducted by Christina Marouli and Cheryl Gomez.

Index

attitude change: common theories of, 56-57; postmodernity and, 60-64

Browning-Ferris Industries, 188 n. 4, 191 nn. 26, 28

Carter administration: implementation of RCRA, 24-27, 137, 168 nn. 21, 23; reaction to Love Canal, 47, 51-52; Superfund initiative, 52, 54, 65-67
Chestnutt, Marty, 159-61, 177-78 n. 12, 178 nn. 16, 17, 192-93 n. 11
Citizen's Clearinghouse for Hazardous Wastes, 72, 74, 76, 81, 82, 93, 98, 152-54
Cohen, Gary, 77, 153
Collette, Will, 70-71, 98, 158, 178-79 n. 19
Commoner, Barry, 27, 32, 82, 153, 170 n. 46
Comprehensive Environmental Response, Compensation and Liability Act (CERCLA). *See* Superfund
Conservation Foundation, 76, 179 n. 31
contamination protest, 41, 84-85, 180 n. 45

Dunn, Lew, 94-97, 156-58, 177-78 n. 12, 178 n. 15

EcoPopulism. *See* radical environmental populism
Environmental Action, 168 n. 8
Environmental Defense Fund, 48, 76, 179 n. 31
environmental equity. *See* race and environment; social class and environment
environmental health and safety

regulations: Clean Air Act, 116, 120; Clean Water Act, 116, 120; Consumer Product Safety Commission, 16, 116, 121, 128; Mine Enforcement and Safety Act, 16, 116; National Environmental Policy Act, 27; National Highway Traffic and Safety Administration, 16, 116; Occupational Safety and Health Act, 16, 116, 121, 128, 129; Resource Recovery Act, 16; Toxic Substances and Control Act, 17. *See also* Hazardous and Solid Waste Amendments; regulatory law, social science studies of; Resource Conservation and Recovery Act; Superfund; Superfund Amendments and Reauthorization Act; U. S. Environmental Protection Agency.
environmental movement. *See* grass-roots environmentalism; mainstream environmentalism
environmental racism. *See* race and environment

Friends of the Earth, 168 n. 8

Gibbs, Lois, 70, 71, 74, 90, 98, 153, 158-60, 192 n. 5, 193 n.12
Gore, Senator Albert, 48, 54, 175 n. 56, 169 n. 25
Gorsuch, Ann, 121, 124, 126, 127, 130, 133
grass-roots environmentalism: growth of, 72-74, 177 nn. 8, 9; ideological development, 77-83; infrastructural development, 74-76, 178-79 n. 19; issue expansion, 76-77; after Love Canal, 70-99; in the 1970s, 69-70; and source reduction, 82, 145-46,

213

149. *See also* contamination protest; movement for environmental justice; radical environmental populism; radicalization; siting opposition
_____: key organizations. *See* Citizen's Clearinghouse for Hazardous Wastes; Greenpeace; National Toxics Campaign
_____: leaders and organizers. *See* Chestnutt, Marty; Cohen, Gary; Collette, Will; Dunn, Lew; Gibbs, Lois; Greer, Sue; Kiker, Kaye; Montague, Peter; Teets, Sally
Greenpeace, 72, 76, 82, 94, 156
Greer, Sue, 70, 71, 77, 80, 90-91, 154-55, 160, 177 nn. 8, 9, 177-78 n. 12, 192 n. 6

Hazardous and Solid Waste Amendments (HSWA): final provisions, 132; implementation, 138; legislative history, 131-32; momentum for passage, 130-31, 187-88 n. 50; proposed provisions, 131; and source reduction, 140, 142
hazardous waste: amounts generated, 11-12, 47; cost of disposal, 12, 144, 145; health effects, 13, 43-47, 52-53, 170 n. 40; improved treatment and disposal methods, 139-40; subpar treatment and disposal practices, 12-13, 36-37, 44, 45-46, 47, 53, 139, 148-49, 171 nn. 58, 59, 191 nn. 26, 28
_____: popular attitudes toward. *See* popular attitudes and beliefs; print media and toxic waste; television news
hazardous waste legislation: before 1976, 11; state laws, 11, 140, 169 n. 25. *See also* Hazardous and Solid Waste Amendments; Resource Conservation and Recovery Act; Superfund; Superfund Amendments and Reauthorization Act

icon: political icon defined, 62-64; toxic waste as icon, 67-68, 125, 128-29, 166
industry—waste generators: actions to impede RCRA implementation, 26; attitude toward generator liability, 20-21, 168 n. 13; attitude toward source reduction, 18-20, 142, 168 n.

12; automakers, 18, 118; campaign for deregulation, 117-19; chemical industry, 11, 12, 19, 20, 21, 42, 43, 45, 46, 48, 65, 72, 79, 118, 141, 142, 174 n. 54, 175-76 n. 57; dislike of regulation, 12, 29, 31-33, 34, 116-17; iron and steel industry, 18, 72; oil industry, 19, 118; plastics industry, 19-20; positions taken on RCRA legislation, 18-21; on siting opposition, 72
_____: disposal practices. *See* hazardous wastes, subpar treatment and disposal practices
industry—waste treatment and disposal: attitudes toward siting opposition and NIMBYism, 73, 78, 80; capacity, 114-15, 182 n. 2; improvement of, 139-40; poor performance standards of, 36-37, 139, 148-49, 171 nn. 58, 59, 191 nn. 26, 28; pro-regulatory statements, 25, 138-39

Kiker, Kaye, 91-94, 98, 155-56, 192 n. 6

Lavelle, Rita, 121, 125-27, 129-30, 187 n. 48
Love Canal: and government officials' perception of the toxic waste issue, 25, 51-54, 122, 124, 125, 129, 131; and popular beliefs, 54-55; in print media, 49-51, 52; and surge of grass-roots activism, 70-71, 104; on television, 42-44, 46-47, 49

mainstream environmentalism: against deregulation, 124; and other political causes, 150; and popular beliefs, 38-39; post-Love Canal line on industrial wastes, 76, 179 n. 31; pre-Love Canal line on waste issues, 14, 35, 168 n. 8; and regulatory laws, 16, 27
_____: key organizations. *See* Conservation Foundation; Environmental Action; Environmental Defense Fund; Friends of the Earth; National Audubon Society; National Wildlife Federation; Natural Resources Defense Council; Sierra Club
media and toxic waste: film, 55-56; and popular beliefs, 39

_____: popular magazines. *See* print media and toxic waste

_____: television. *See* television news

Montague, Peter, 77, 153

movement for environmental justice, 150-61; redefinition of environmentalism, 151, 152; solidarity with other movements, 150, 152-53. *See also* race and environment; radical environmental populism; radicalization; women and toxic waste

National Audubon Society, 76, 179 n. 31

National Toxics Campaign, 72, 75, 77, 81, 82, 91, 97, 98, 153, 155

National Wildlife Federation, 76, 179 n. 31

Natural Resources Defense Council, 76, 179 n. 31

Nixon administration and waste policy, 16-17, 22

not in my backyard (NIMBY): condemnation of, 77-80, 180 n. 35; falseness of charges of, 80-83, 146-49; truth of, 80, 82. *See also* siting opposition

pollution prevention. *See* source reduction

popular attitudes and beliefs: about deregulation, 117, 119-20, 128; distrust of industry, 36-37, 39, 55, 78, 87, 104-5, 148-49; distrust of regulators, 36, 86, 87, 104-5, 148-49; about environmental problems, 39; about hazardous waste, 13-15, 54-55; problematic salience of, 39-41, 63-64, 67-68; about technology, 38, 104, 146. *See also* attitude change

postmodernity: theories of, 57-60, 173 n. 38, 173-74 n. 40

_____: and attitude change. *See* icon: political icon defined

print media and toxic waste, 49-51, 172 n. 19

public opinion. *See* popular attitudes and beliefs

race and environment: activism by people of color, 75-76, 151; connection between environmentalism and struggle for racial justice, 152-53; differential toxic burden, 75, 147, 151;

environmentalism redefined, 151-52, 192 n. 7; racial tensions in grass-roots activism, 192 n. 6

radical environmental populism, 77-83, 89-99

radicalization: individual-level, 90-97, 153-61; movement organization-level, 89-90, 97-99, 150-53

Reagan administration: concessions after Sewergate, 132-33; deregulation of RCRA and Superfund, 87, 104, 121-23, 125; deregulatory policy, in general, 120-21

_____: EPA scandal during. *See* Sewergate

regulatory law, social science studies of, 28-34, 169 n. 30, 170 n. 43

Resource Conservation and Recovery Act (RCRA): implementation, 23-27, 87, 104, 121-22, 125, 129; legislative history, 16-23; provisions, 23, 110; and source reduction, 17-22

_____: reauthorization. *See* Hazardous and Solid Waste Amendments

risk perception. *See* attitude change; popular attitudes and beliefs

Sewergate, 87, 88, 123-30

Sierra Club, 76, 179 n. 31

siting opposition: causes of, 41, 85-86, 146-49; conditions favoring, 86-88, 181 n. 52; labeled a problem, 36, 41, 69-70, 71, 72-74, 103-5, 146, 178 n. 14. *See also* grass-roots environmentalism; siting strategies

siting strategies: compensation, 108-10; expanded participation, 110-13; failure of, 114-15, 183 nn. 26, 27, 184 n. 28; finding isolated sites, 105-6; finding weak or amenable communities, 106-7; preemption, 107-8; state siting statutes, 113-14

social class and environment: differential toxic burden, 147, 151; jobs versus environment, 18-19, 168 n. 10

social movements theory: social movement emergence, 84-88; social movement evolution, 88-90, 97-99

source reduction, 17-22, 140-46; causes of the shift toward, 141, 142-46, 149, 189-90 nn. 21, 22; grass-roots environmentalism and, 82, 145-46,

149; industry attitudes toward, 18-20, 142; official policy, 21-22, 140-42, 189 n. 12

Stockman, David, 65, 120, 176 n. 58, 185 n. 10

Superfund: deregulation and, 122-23; implementation, 122-23, 129-30, 138-39; legislative history, 65-67, 175-76 nn. 55, 56, 57; and Love Canal, 53-54, 63-64; provisions, 66-67, 144

_____: reauthorization. *See* Superfund Amendments and Reauthorization Act

Superfund Amendments and Reauthorization Act: final provisions, 133, 145; implementation, 138-39; legislative history, 132-33, 188 n. 56; momentum for passage, 133; proposed provisions, 132-33

Teets, Sally, 160, 177 nn. 8, 9, 178 n. 17

television news: coverage of environmental activism, 44, 46-47, 67; coverage of environmental problems, 1-3, 39; coverage of hazardous wastes, 42-49; social science studies of, 41-42

toxic waste. *See* hazardous waste

toxics movement. *See* grass-roots environmentalism

U.S. Congress: and deregulation, 123-28; post-Love Canal interest in toxic waste,

51-54, 129; RCRA implementation oversight, 24-26; RCRA legislative history, 16-23; RCRA reauthorization, 131-32; on source reduction, 17-22, 140, 142; Superfund legislation, 65-67, 174-76 nn. 54-59; Superfund reauthorization, 132-33

U.S. Environmental Protection Agency (EPA): deregulation, in Reagan administration, 121-23, 125; implementation of RCRA in Carter administration, 24-27, 184-85 n. 9; reactions to Love Canal, 44, 46-47; on siting opposition and NIMBYism, 71, 73, 78, 79; on siting strategies, 108, 111-12, 113, 114; and source reduction, 140, 142, 143; statements about hazardous waste problem, 12-13, 44, 46

_____: 1983 scandal. *See* Sewergate

Waste Management, Inc., 91, 93, 188 n. 4, 191 n. 28

waste minimization. *See* source reduction

women and toxic waste: fighting sexism part of grass-roots activism, 152; solidarity between women's movement and environmental movement, 150, 153, 192 nn. 4, 8; women as movement leaders and cadre, 152, 192 n. 8. *See also* movement for environmental justice; radical environmental populism

Andrew Szasz received his Ph.D. in sociology from the University of Wisconsin—Madison in 1982. He is an associate professor in the Board of Studies in Sociology at the University of California at Santa Cruz. He teaches courses on environmental sociology and on social theory, and is organizing a graduate program in environmental sociology. At UCSC, Dr. Szasz is also affiliated with the Board of Studies in Environmental Studies and with the Center for the Study of Global Transformation. He has served on the governing council of the Environment and Technology Section of the American Sociological Association. Before his current research on the politics of toxic waste, Dr. Szasz studied occupational safety and health policy in the United States. He has published numerous articles on regulation, worker safety and health, and hazardous waste.